Carnival Art, Culture and Politics

Drawing on rich insights from cultural, post-structural and post-colonial studies, this book unlike any other, demands that we re-think Carnival and the carnivalesque as not just celebratory moments or even as critical subtext, but also as insightful performatives of social life anywhere, given the entangled times and spaces of these performances. The authors review Carnival's performative aspects not merely as a calendrical festival, but rather center attention on the relationship between carnival and everyday life, and on how people negotiate their social spaces and possibilities in the context of modern power. The book therefore seeks to highlight the knotted time-spaces of power and to demonstrate the dynamic interplay between state spaces and people's spaces that are being weaved by carnival's interlocutors. It demonstrates how Carnival and the carnivalesque become analytic optics through which the relations of power in the social and political life of subjects who seek to tactically or strategically vary their given identities, can be productively engaged.

This book was originally published as a special issue of *Social Identities: Journal for the Study of Race, Nation and Culture.*

Michaeline A. Crichlow is a Professor of Sociology and Global Caribbean Studies, teaches in the African and African American Studies department at Duke University, USA, and is the author of *Globalization and the Post-Creole Imagination: Notes on Fleeing the Plantation* (2009) and *Negotiating Caribbean Freedom: Peasants and the State in Development* (2005). She has written and published on informality, creolization and development in several journals and edited volumes, and is currently researching for book projects on citizenship and migration and the question of place and space.

Carnival Art, Culture and Politics

Performing Life

Edited by

Michaeline A. Crichlow

Routledge
Taylor & Francis Group
LONDON AND NEW YORK

First published 2012
by Routledge
4 Park Square, Milton Park, Abingdon, Oxon OX14 4RN
605 Third Avenue, New York, NY 10017

Routledge is an imprint of the Taylor & Francis Group, an informa business

This book is a reproduction of *Social Identities: Journal for the Study of Race, Nation and Culture*, volume 16/issue 4. The Publisher requests to those authors who may be citing this book to state, also, the bibliographical details of the special issue on which the book was based.

British Library Cataloguing in Publication Data
A catalogue record for this book is available from the British Library

ISBN13: 978-0-415-69276-2 (hbk)

Typeset in Times New Roman
by Taylor & Francis Books

Publisher's Note
The publisher would like to make readers aware that the chapters in this book may be referred to as articles as they are identical to the articles published in the special issue. The publisher accepts responsibility for any inconsistencies that may have arisen in the course of preparing this volume for print.

Contents

INTRODUCTION

Carnival praxis, carnivalesque strategies and Atlantic interstices

Michaeline A. Crichlow[a] and Piers Armstrong[b]

[a]*Departments of African and African American Studies & Sociology, Duke University, Durham, North Carolina;* [b]*Latin American Institute, University of California, Los Angeles; Modern Languages and Literatures, California State University, Los Angeles*

Somo crioulo doido e somo bem legal
Temos cabelo duro é só no black pau
Que bloco é esse? Eu quero saber.
É o mundo negro que viemos mostrar pra você.
Branco, se você soubesse o valor que o preto tem,
Tu tomavas banho de piche, branco e, ficava negrão também.
E não te ensino a minha malandragem
Nem tão pouco minha filosofia, não

We be crazy Creole and we be very cool
We have kinky hair and we just go in 'Black Pow'
What carnival club is this? I want to know.
It's the black world that we came to show you
Whitey, if you knew the value of the Black man,
You'd bathe yourself in tar and become a Black man too
And I'm not gonna teach you my artful dodging
or my philosophy either, no. (Que Bloco É Esse, Paulinho Camafeu)

A multidisciplinary symposium on carnivals and the carnivalesque

In the summer of 2005, 11 scholars who had studied and participated in carnival and carnivalesque moments, gathered at the Obermann Center for Advanced Studies at the University of Iowa in Iowa City for a colloquium on carnival. Their respective core disciplines included sociology, anthropology, history, folklore, visual arts, theatre, performance, and foreign languages and literatures. The carnivals they studied were predominantly of the Americas (with a Caribbean cluster of Trinidad and Tobago, Cuba and New Orleans, and an Eastern South American cluster of Brazil, Argentina and Uruguay) but also included France, the Trinidadian-derived Notting Hill carnival in London, a carnival-like Catholic festivity in the Philippines, and on the carnivalesque front a G-8 protest in Scotland and well-humoured coal industry protests in West Virginia. All used interdisciplinary perspectives and methods in their work, and the colloquium fostered exchanges of a comparative nature regarding both instances of carnival and its disciplinary frames. The conceptual orientation deriving from the profile of case studies examined, meant combining a focus on post-colonial issues of emancipation with aesthetic performatives and the expression of personal fantasies. We sought to explore carnival beyond its traditional festive sites, and to rethink it, less through overarching binaries such as Western/non-Western or domination/subordination, and more in terms of its cultural flows, hybridities, negotiation, dialogics, and unintended consequences.

An inclination to rely on binaristic notions in earlier carnival literature, particularly within Marxist and certain ethnographic studies, tended to underscore the 'grotesque' anomalies in a visual politics of presence or place. But the cultural turn in social studies, with its emphasis on translation, negotiation, constructions of subjectivities, and on liminality, allows more for a perspective of open-ended transformations and relational histories, and hence a stress upon the contingent, provisional formation of social spaces, subjectivities and temporalities. This view's value has been enhanced with the recent overflow of certain carnivals beyond their historical sites and temporal constraints. The essays in this volume offer then to rethink carnivals and the carnivalesque principally in terms of its movements, boundary crossings, convivialities, creative bricolage and fantasies, which are maneuvers open to anyone – state elites, 'folk' and visitors, players and observers.

Carnival *per se*

Carnival was originally a Catholic festival, lasting for three–five days, just prior to the 40 days of Lent which precede Easter and are marked by a penitential restraint which is empathic to Christ's impending agony. The etymology of the word is contested, but apparently derives from the Italian, *carne levare* (the removal of meat). Since it has not traditionally existed in Anglophone societies, there is some awkwardness in English usage of the term. 'Carnival' usually refers to a temporary amusement park, and Americans use a French term in New Orleans, *mardi gras* (Fat Tuesday), when people would eat up all the meat and fat proscribed in Lent, which starts the following day, Ash Wednesday, the day of ritual mourning that follows carnival. Since Lent was a time of penance and discipline, the preceding days became, in contrast, a time of liberation from the usual social and moral constraints, a public party. In medieval times, carnival was part of an organic cycle of discipline and liberation. For a day the fool or the fattest glutton in the town became 'king', and, to a lesser extent, or at least by implication, the 'king' (i.e., the local baron or the mayor or other leading burgher) became a 'fool'. Usually, instead of leading to actual liberation, carnival was politically useful to the powerful as a harmless escape valve for oppressed people. It was, nevertheless, predicated psycho-socially on a symbolic inversion of status which is highly ambiguous: if carnival helped to maintain the status quo, it uniquely allowed for bold expressions of dissidence and thus pointed to a 'revolution in the mind' in the form of a kind of alternative reality. Carnival was also singular in precipitating a gregarious, compacted assemblage of townsfolk of different stations in a general clime which embraced both meddling with order and individual eccentricity. Finally, at a deeper spiritual level, carnival was profoundly enigmatic in that it preserved links with pagan practices and sensibilities, and presumably initially emerged as a political accommodation, by the Church, of enduring pre-Christian values. From this point of view, carnival is simply a mask for the continuity of Bacchic rites which are traced back to Roman festivities and, beyond that, to pagan principles of the earlier ancient world.

In the early modern historical period (after the Middle Ages and before the Industrial and French revolutions), the enigmatic ritual aspects of carnival receded with economic development. Given the political volatility of this period of national consolidation, carnival became an arena for expressions of changing patronal power (for example, when a powerful burgher sought to mark his ascension in a town). The line that carnival constantly sketches between the potential and the real was sometimes broken, and the margin of unruliness tolerated in carnival occasionally broke into purposeful violence because of the links to extraneous political agendas. Emmanuel LeRoy Ladurie's rigorous study (1979) demonstrates how carnival was used as a venue to channel hostility deriving from anger over taxation and corruption in the linguistically divided southern French town of Romans, culminating in widespread violence in 1580 and the massacre of two dozen artisans. In a different vein, Teófilo Ruiz (2001) documents the consistency of non-cataclysmic manipulations of carnival for political assertion and for internecine or inter-class intimidation. In the later modern period in Europe, carnival generally lost both its ritual and political edges and evolved into placid communitarian celebration, aesthetic stylization and commodification. The most patent manifestation of this is the carnival of Venice. Largely because of tourism, this carnival has remained illustrious despite the fading away of the economics and politics which saw its rise. In a short but brilliant study, Peter Burke (1987) traces its gradual evolution through the early modern period. What now appears as the most stable carnival, with a sacrosanct aesthetic transposition (the use of *ancien regime* masks), was in fact constantly adjusting, both logistically and aesthetically, in reaction to changing power relations.

During this relative metropolitan decline, and particularly in the twentieth century, the carnival became gradually more important in the non-Anglophone colonies and former colonies of the Americas (based on its distinct colonial history, we can include New Orleans in the non-Anglophone Caribbean). While Rio's is the most famous, Brazil has several major and distinct carnivals, and the carnival is the longest official holiday period of the year. Lesser known carnivals are central national events in Caribbean states such as Haiti, Cuba and Martinique. Trinidad (i.e., Trinidad and Tobago) occupies a special place at the carnival cross-roads. Colonized first by the Spanish and the French, it is perched with the previous States at the Afro-Latin fulcrum of New World carnival. Eventually an Anglophone colony, it was also a full participant in the vibrant popular cultures of the British Commonwealth, in which, particularly in the mid-twentieth century, there were often thematic overlaps between political and cultural independence. In Trinidad, carnival is the central arena of popular cultural articulations. Additionally, Trinidadian culture and carnival are infused with East Indian culture, thus going beyond the usual concepts of the 'Black Atlantic' or 'trans-Atlantic'. This global triangulation should be distinguished both from the Afro-Latin core of most Caribbean and Brazilian carnivals, and from the *mestizo* Afro-indigenous interfaces of Spanish America which have their own carnival expressions (in Mexico, Bolivia, Peru, etc.). The New Orleans and the Recife (Brazil) carnivals feature a rich Black assimilation of indigenous totems. The Rio carnival functions as a spectacular ecumenical ritual of national inter-ethnic unification, and clearly as an escape valve.

The Bahian carnival (also Brazil), meanwhile, has been a central site of Black rejection of the 'racial democracy' ethos articulated in Rio. While the Bahian and Trinidadian carnival constantly evolve, it is revealing that the enduring format of the Rio carnival was established during a quasi-Fascist dictatorship in the 1930s. In fact, the populist political dimension of New World carnival is constant. For example, Scher (2002) examines the deployment of carnival for state purposes in Trinidad and Averill (1994) and Largey (2000) have shown how the contemporary carnival of Haiti (*Rara*) serves to consolidate state power on a cruder level. This process, further radicalized, is elucidated by emerging African theories of the carnivalization of state power in that continent's post-colonial states (Mbembe, 1992, 2001). In the future, these theories of the carnivalesque may illuminate analyses of Middle Eastern cultural mechanics which problematize the Western project of hurriedly imposing democratization.

Theories

> The dusk was a raucous chaos of curses, gossip and laughter, everything performed in public... [To] colonials... the self-inflicted role of martyr came naturally, the melodramatic belief that one was the message-bearer for the millennium... Centuries of servitude have to be shucked; but there is no history, only the history of emotion. Pubescent ignorance comes into the light, a shy girl, eager to charm, and one's instinct is savage: to violate that ingenuousness... because she too moves in her own hallucination: that of a fine and separate star, while her counterpart, the actor, sits watching, but he sits next to another hallucination, a doppelganger released from his environment and his race. Their simplicity is really ambition. Their gaze is filmed with hope of departure. The noblest are those who are trapped, who have accepted the twilight.... Every state sees its image in those forms which have the mass appeal of sport, seasonal and amateurish. Stamped on that image is the old colonial grimace of the laughing nigger, steelbandsman, carnival masker, calypsonian and limbo dancer. These popular artists are trapped in the State's concept of the folk form, for they preserve the colonial demeanour and threaten nothing. The folk arts have become the symbol of a carefree, accommodating culture, an adjunct to tourism... (Derek Walcott, 1970)

Scholars vary in focus and concern as they read political, social and aesthetic meanings and functions, on the collective and individual planes. Given the complexity of this phenomena, what are we looking for on the ground? First, carnival in a postcolonial frame precipitates questions including the racial condition in the New World, its slave post-slavery and indentured labour contexts, of its contestations and cultural politics in reaction to the increasing social and cultural heterogeneity of Europe, hybrid praxes and identities, and the question of 'roots and routes'. Second, carnival precipitates political and economic questions, on the one hand with a view to scrutinizing (top-down) state interventions and impositions, and, on the other, with bottom-up or horizontal carnivalesque subversions and general manoeuvres. By horizontal, we mean to eschew binarisms of race, class, gender and state that wittingly or unwittingly convey power as lived through structural determinations and thereby the cosmovision of dominant groups and identities. Third, following Bourdieu and his key notion of socio-cultural domains ('habitus'), we should be attentive to the interaction and mutual influence between carnival performance genres and sociabilities and the power matrices that inform them:

> [T]he states or social contexts associated with a particular class of conditions of existence produce *habitus*, systems of durable, transposable dispositions, *structured structures predisposed to function as structuring structures*, that is, as principles which generate and organize practices and representations that can be objectively adapted to their outcomes without presupposing a conscious aiming at ends or an express mastery of the operations necessary in order to attain them. (Bourdieu, 1990, pp. 53–56)

Fourth, while performance may imply politics and/or be a reaction to one's social position, it manifests aesthetically and warrants analysis as such. Finally, while carnival in its classic form is scheduled and anticipated, the carnivalesque refers to things which can be said to 'happen', whether in spontaneous outbreaks of carnival-like practices, or in strategic actions inspired by carnival. The carnivalesque may also emerge organically from traditional practices as something new (and thereupon possibly be integrated into a recast of the 'traditional'), or it may be extracted from external sources in the global imaginary.

Carnival studies begin with Bakhtin and since attempts to go beyond still tend to go back to him, the result is that it is never clear whether the Russian literary philosopher is one part of carnival theory or, rather, the reverse. Bakhtin's seminal ideas of carnival were framed as an introduction to a doctoral thesis on Rabelais (submitted in 1941, published in 1965, translated to English in 1968). His interest is more in the carnivalesque – for him, bottom-up societal propulsions to dissidence and pluralism which manifest in carnival, and the complementary nature of this moment to normal time and order – than in carnival ethnographies and *in situ* studies *per se*. Since Bakhtin lived under Stalin, it is difficult not to read his recuperation of the late medieval popular as a couched reaction and a symbolic counterpoint to early Marxist dismissals of the popular as alienation. This only exacerbates the lack of ethnographic groundedness, especially if contrasted with the great socio-anthropological studies of popular culture which emerged from the Americas in the twentieth century (such as Boas and later Geertz in the USA, Freyre in Brazil and Ortiz in Cuba), European social historicism (such as Eric Hobsbawm, and E.P. Thompson), later cultural studies of New Historicism (such as Greenblatt), and various recent African, Latin American and Asian theorists of cultural particularism (such as Thiong'o, Said and Canclini). Bakhtin's notion of the carnivalesque itself fits into a panoply of his other larger ideas about the dialogical relations of discourses in a society. This entails the mapping of characteristic discourses to subject-positions (or the societal insertions of the social sub-sets that make up the whole).

This sophisticated understanding of pluralism, whose dimensions go beyond conventional political representations, has afforded Bakhtinian theory (beyond carnival) importance to new theorists of radical democracy in the post-Marxian moment. As Hirschkop and Shepherd (2001) outline, while various post-structuralists appropriated Bakhtin in the 1970s, Stallybrass and White (1986) countered this highly abstract trend with more socio-historic precision, a scepticism as to the notion of Bakhtin as a Marxist thinker and yet a validation of his relevance to interventionist Leftist agendas. Hitchcock (1993) is emblematic of the application of Bakhtin to the subaltern (in this case, the female subaltern, in literature). Hitchcock's critical reading of Bakhtin's notion of dialogism suggests a framework that encompasses a wide range of contestatory practices that not only reflect the powers in the social, but also operate as a medium of refraction – 'a mediation of social conflict at any one moment in history' (Hitchcock, 1993, p. 4). But Hitchcock points out that though Bakhtin stays close to an idea of the struggle of 'the oppressed', there is no guarantee or necessary expectation for social transformation to be realized simply because of the emergence and expressions of performative plays with and through 'Power'. In his own theoretical Bakhtinian ventures, Hirschkop (1999) proceeds cautiously but hopefully in making Bakhtin pertain to politics. This leads him back to the literary because it was the central subject of Bakhtin, if not his ultimate concern. Hirschkop reminds us of the linguistic arena and Bakhtin's insistence on the social interrelatedness of utterances versus Saussure's neat structuralist exposition of the sign. If one is to nominate a discipline for Bakhtin, it might well be Communication Studies, but only with the caveat that Bakhtin moves between the highest existential intimations (communication as the quasi-spiritual) and the mechanical precision and literary sensitivity of formalism. Neither the abstract post-structuralist appropriations, nor the cultural political theorists (e.g. Hirschkop), nor the applied cultural studies theorists who use Bakhtin so effectively (e.g. Stam, 1989), nor the literary scholars who explore Bakhtin on his own terms (Emerson, 1997; Clark & Holquist, 1984) are of obvious use to scholars of actual carnivals, since the latter's studies almost inevitably involve grounded ethnographic knowledge (one exception to this is Roberto DaMatta, who masterfully applies Bakhtinian schemas of inversion to his reading of Rio's carnival and New Orleans' *mardi gras*; DaMatta also draws, however, on anthropological theorists of symbolic inversion; see DaMatta, 1991).

The most imposing Bakhtinian elements in carnival studies are the core categories of the mystery of carnival: carnival and ordinary time, or carnival as a twilight zone of possibility; a repressed sovereign popular will in opposition to conventional authority; the inversion of the usual totemic and taboo status of certain social stations and ethnic profiles; the tantalizing role of the grotesque (or socially stigmatized) in this inversion; finally, the enigma of carnival's poise between its license for dissidence and its totalizing coalescing force. The anthropologist Victor Turner's work (1969) on 'liminal' psycho-social states during cathartic festivals (including Rio's carnival; see Turner, 1987) is a rare instance of an ethnographically grounded refinement of Bakhtin's insights.

On a more basic level, carnival is affected by Bakhtin's anointing of the popular *per se*. Countless achievements and perceptions have consolidated the new prestige of the popular in the twentieth century, from the reception of primitive art in modernism and subsequent assailments on high culture, through the youth cultural revolution's music-centeredness and on to the resuscitation of poetry in oral rap. Bakhtin's celebration of carnival and his other theories have served as powerful validation of oral and performative popular culture as distinct from literary culture *per se*, and prodded our awareness that the deepest culture is informally acquired cultural practice rather than either high or commodified low culture.

Bakhtin's premonitory illumination of subaltern potentiality remains important to very distinct theoretical trajectories. The European scholarship outlined above constitutes a high literary philosophy which follows deconstruction and post-structuralism and interrogates cultural ontology at a rather abstract level while still being committed to subaltern particularities. Tzvetan Todorov, whose work is followed and disseminated in the Bakhtin Centre at the University of Sheffield, has worked extensively on the European/non-European interface in the

Americas (1982/1984), though with a focus on the indigenous rather than the Black or Creole Other. Paul Gilroy's pioneering work (1993) in the 'Black Atlantic' integrates Bakhtin to a degree. His recast of the notion of double-consciousness – incipient in Bakhtin but more importantly prevalent in the great twentieth-century Protestant Black American literary heritage of Du Bois, Ellison and others, and central in now-dominant theories of identity politics – celebrates Black Atlantic aesthetic achievement built on a synthetic consciousness grappling contradictions (particularly in popular music). Gilroy's sense of the need to denounce political contradiction, notably the universalist illusions of the Euro-centric enlightenment, and to materially galvanize subalterity into political self-consciousness, sits ill at ease with Afro-Latin ecumenicalism, whether in the form of Freyre's apologia for racial democracy (1933/1963) or its popular correlates in New World carnivals – in short, with carnival as a positive (Coleridge) rather than a negative opiate (Marx).

Insofar as New World carnival points to race, it does so in ways as complexified as New World racio-cultural hybridities themselves. This proliferation of differences makes for a continuum of identities from larger to smaller groups and finally to the individual (and we should keep in mind here the Latin American propensity to subtle distinctions in the nomenclature for racial mixes, in contrast with the binarism of the US). This parallels a continuum which is constant in carnival, between projections and experiences which are collective and those which mark an individual as a differentiated entity. At this point – where binaries fragment – performance theories can dextrously abet our tracing of the surplus meanings of carnival. In dialogue with the postcolonial and other fields, performance studies can bridge the ludic and the social, with particular attention to the subaltern – the gay, the lesbian, the woman, the Black or the White, who violates *de rigueur* solidarity to pursue Whiteness or Blackness or some other Otherness. New World carnival also often points to sexual personae – indeed, without analysing further here for lack of space, it is tantalizing that New World carnivals and their societies are radically more carnivalesque than their Old World correlates as regards gender, while this relation is inversed as regards orthodox gender liberation in the everyday order. Perhaps carnival is to the everyday, as drag is to heterosexuality. For Judith Butler, 'drag does not oppose heterosexuality', nor does it automatically subvert it.

> [O]n the contrary, drag tends to be the allegorization of heterosexuality and its constitutive melancholia. As an allegory that works through the hyperbolic, drag brings into relief what is, after all, determined only in relation to the hyperbolic: the understated, taken-for-granted quality of heterosexual performativity. At its best, then, drag can be read for the way in which hyperbolic norms are dissimulated as the heterosexual mundane. (Butler, 1993, p. 27)

Evidently, Butler's comments on gender can also speak to the ethnically-inflected self-recasts common in carnival play, particularly those by solo individuals. Her idea of the melancholia of orthodoxy also speaks to why subalterns tend to emerge as the *raison d'être* of New World carnivals – not only because of the opportunities the event affords subalterns for expression, but also for the needs and fantasies of the society – including its everyday masters, as they gaze on and participate in the event. But carnival suggests even more about this melancholia: Derek Walcott (in the epigraph above) pessimistically points to the melancholia not of the master but of the postcolonial subaltern, if ever s/he should stop to look at her/his own shadow and think. Just as carnival is about multifarious projections of concocted identities, it is about running from one's own shadow, never looking back for fear of turning into a pillar of salt. But grasped in all of its fullness, its participants escape the 'false consciousness', that Walcott intimates. Conversant with all these perspectives, Aching (2002) develops a persuasive reading of the Caribbean which takes the carnival motif of masking not as a strategy reserved for carnival time but as a general existential *modus operandi* in the Creole world. This focus on an active social intercourse which is subjectively deployed contrasts starkly with Mbembe's postcolonial writing on Africa. Mbembe (1992) uses Bakhtin's notion of 'grotesque realism' and the carnivalesque, but argues that they are entangled with state crafted and state perpetrated phenomena. Popular subjectivity is seen as trapped within this state-commandeered register of banal power. Worse, state and people end in a collusion whereby the tyrant proceeds with the tacit complicity, mimicry, and conviviality of the African subject.

In fact, carnival and the carnivalesque are double-edged rather than monolithic. As is performance. Drewal considers that Butler ends up 'reading insurrection as either misappropriation or expropriation' (Drewal, 2000, p. 117); Drewal revalidates

> the contestatory capacity of performers to embody skills and techniques completely alien to normalized social conventions, and an active remembering of that forgotten history that unmoors performance and performers from constrained contingency and liberates the body for insurrection without recourse to either misappropriation or expropriation. (p. 119)

As usual, carnival seen at ground-zero affirms and complicates this, reiterating boundaries as much as breaking them. Carnival play is not fully euphoric, liberatory or anarchic. In terms of gender-crossing, for example, Real (2004) documents the performance of drag in Brazil's 'Desfile dos Virgens do Bairro Novo' (Parade of the 'Virgins' of the Barrio Novo suburb) in Olinda during pre-carnival week, by ostensibly heterosexual men, who in performing their gender caricatures exclude gays and act more in the mocking spirit of blackface than in solidarity with transvestites.

This brings us back to that aspect of performance which concerns the collective pole of carnival. Here, genres are the refraction of social histories as channelled by aesthetic traditions, and the embodiment of identities; these multiple ideas are synthetically encapsulated in the term 'performative histories' (Crichlow 2009). How far back do or should these histories reach? In the case of Amerindian and *mestizo* carnivals and other festivities, scholars routinely posit links back to pre-Columbian roots. Because of the traumatic preponderant weight of slavery and post-slavery, North Atlantic socio-anthropological scholarship on Blacks in the New World has generally been less interested in the African than the Black, and, because of the residues of racism, wary of cultural essentialism. Esiaba Irobi (2007), however, contests Gilroy's (and others') negligence of continental African influences on New World carnival, arguing for the continuity of 'performative literacies' grounded not only in aesthetic praxes *per se* but more generally in subjective social intercourse. Irobi demonstrates this by comparing West African festival theatre with various carnivals of the African diaspora, including Trinidad and Notting Hill. In dealing with the body – and embodied memory – Irobi characterizes the Bakhtinian emphasis on the scatological as a European view tinged with Christian notions of purgation, and one which quite misses the constructive, positive and non-climactic essence of West African public festivities which serve as a site for ancestral spirit evocation and incorporation. In Irobi's perspective then, carnival is portrayed as presenting an African influence on a European tradition when the process should really be imagined in reverse, as an assimilation of European venues and banners into a dynamic, evolving African sensibility which is not constrained in the mother continent but rather flows out to its diaspora and into the wider world. Notwithstanding the validity of Irobi's ethnographic corrective as to carnival, this says little about postmodern, wandering invocations of the carnivalesque, such as the studies of Bogad and Hufford in this volume. Irobi's views point to a prospective phenomenon broader than the various provincialisms of Bakhtin's Euro-Christian carnival, Irobi's own West African festive theatre, the apologists of Ibero-American racial democracy, the new identity politics theorists, and his unwitting recentering of the West/non-West binary. This broader phenomenon would be the carnivalesque as a universal category, somehow as integral to the human psyche and to culture as art or laughter or Oedipus.

Given this range of takes, our survey of theoretical perspectives simply opens up more questions. How do these performances preserve or rupture traditions? How are they spectacularized in public performances? What sorts of incomplete politics, or poetics, are being orchestrated here? What dissidence occurs and what is merely rehearsed? Are there returns to the centre which re-align marginalities? More simply, to what extent is carnival the inverse or the reflection of everyday order? Clearly, carnival events precipitate a politics of place and space as the enactments of difference – but what is the 'proper' taxonomy of difference, and what, in the end, are the political or existential meanings of 'enactment?'

Practices

Understandably, most theoretical discussions of carnival disintegrate, or reintegrate, into discussions of concrete instances. The term 'performative histories' alerts us to socio-historic dimensions which are easily comparable between groups, but also to the aesthetic autonomy of performance. The latter concept also morphs, in discussion, from the abstract to the concrete, whether in the discernment of the *language* of a given genre, or casual observation of stylistic intangibles and quirks. A brief outline of carnival practices might do well similarly to move from the general to the intricacies of genre and performance.

The first and most global point is that carnival is always kinaesthetic, that is, it involves movement and resists stasis. This enacts (performs and represents) the idea of transformation. The most monumental set pieces (carnival floats, etc.) reveal a transcendent transfiguration, but they must still roll down the street, and eventually recede from view. In its simplest modality, the carnival manifests as a group of merrymakers who must also come and go. Yet visuals, considered as stills rather than kinaesthetically, are an integral feature of carnival. The common practice of masking and wearing costumes means both the projection of some abstract social identity and the deployment of plastic arts. This is the soul of the Venice carnival. In the

Americas, while constantly present, its centrality varies; it is relatively more important in the carnivals of New Orleans and Recife, for example. Changing visuals can be read as reflections of changing social agendas and of changing ethnic interfaces, as Bettelheim (2001) shows for Cuba. The unique flamboyance of the Rio spectacle draws on Amerindian influences in the use of loud colours and feathers in individual costumes, and Western Mediterranean influences (Spain, Italy) in the use of large-scale visuals of the puppet type. The social meanings of these figures, usually in the spirit of *bon homie*, famous personages and the buffoon-as-carnival-king, are clear. Trinidad's large-scale *mas* presentations (*mas* derives from mask and masquerade; *mas* can be a group or individual undertaking) have developed more challenging abstract visuals, particularly in the work of Peter Minshall (see Schechner & Riggio, 1998). As Kerrigan and Laughlin (2004) comment:

> Minshall's bands often address social or metaphysical questions. 1980's *Danse Macabre* was a sardonic death masque. 1982's *Papillon* was a meditation on the fragility of life. In 1983 he began the trilogy that may be his magnum opus with *River*. His masqueraders danced on the Savannah stage in innocent white cloth; then jets of coloured dye suddenly appeared, staining the River People. His *Washerwoman* queen, the embodiment of purity and harmony, was symbolically raped and murdered by his king, *Mancrab* [see Figure 1], a devil-like creature representing greed and evil technology. This manifestation of modern anxiety was followed in 1984 by *Callaloo*, then in 1985 by the final act, *The Golden Calabash*, two bands in one. *The Lords of Light* and *The Princes of Darkness* clashed before the Savannah audience in a supreme battle between good and evil. (Kerrigan & Laughlin, 2004)

In many ways, Rio's and Trinidad's carnivals are pillars of New World carnivals, between which most other carnivals can be situated (see Armstrong elsewhere in this volume). The Rio event is predicated on plastic arts with Hollywood standards, vast human choreographies, extraordinary musical precision by hundreds of percussionists, and logistic discipline. Tellingly, this transpires in a stadium pastiched as a street. Along with Minshall's intensity, arguably Trinidad's carnival is probably the richest in the world in manifestations of individual fantasy and theatricality, small group sorties, large (steel) band performances, and associative musical genres – notably the jousting art of calypso. The remarkable variety of Trinidadian carnival is well documented in Milla Riggio's anthology (2004). In that volume, the ethnomusicologist Gordon Rohlehr (2004) describes both the socio-political acumen expected of calypso, and the musical stylistic evolution into and beyond calypso. In a similar spirit, Armstrong (2001) traces the evolution of Bahian carnival song genres. Very broadly speaking, while New World carnivals are visually innovative, music is the soul of carnival, at least in the Afro-Latin Caribbean and South Atlantic contexts predominant in the present collection of essays. The performative demands of music – it is not easy for outsiders to learn to play or even 'get' many carnival rhythms, much less dance well – tend to enhance the collective pole of carnival and the socio-ethnic anchoring of what we have called performative histories. But we have to insist, once again, on the gregarious nature of carnival. Usually, anyone is welcome to join at least as a fellow-traveller in such practices. In this spirit, it is worth noting that independent of Rohlehr's ethnomusicological erudition, his text is suggestive on another, broader level: in the stage-names of once prominent calypsonians – Mighty Sparrow, Lord Kitchener, Lord Invader, Shadow, Mighty Growler, Lord Shorty, Ras Shorty I – we find again the

Figure 1. *Mancrab, from the Mas' band River*, by Peter Minshall, Trinidad Carnival 1983. Photo by Noel Norton.

carnival proclivity to horizontalize hierarchies with humour which is often self-deprecating.

The essays

> Fonte de mel/Nos olhos de gueixa/Kabuki, máscara... Beleza esperta/Você me deixa a rua deserta/Quando atravessa/E não olha pra trás... Onda do mar do amor/Que bateu em mim/Você é forte/Dentes e músculos/Peitos e lábios/Você é forte/Letras e músicas/ Todas as músicas/Que ainda hei de ouvir /... Gosto de ver/Você no seu ritmo/Dona do carnaval/Gosto de ter/Sentir seu estilo/Ir no seu íntimo/Nunca me faça mal.
>
> Fountain of honey/in Geisha eyes/Kabuki, mask... Clever beauty/You leave me in an empty street/When you cross/and don't look back /... An ocean wave of love/That broke upon me/You are strong/teeth and muscles/Breasts and lips/You are strong/ Lyrics and songs/All the songs/I will ever hear... I like to see/You in your rhythm/ Lady of carnival/I like to hold/To feel your style/To enter into your intimacy/Never do me harm. (Caetano Veloso, 'Você é linda')

In this journal collection we emphasize how carnival enables a space to play out and pose issues of incompleteness, contingencies, fantasies and transitions. Carnivalesque performances can destabilize the sites where they occur and draw attention to the imagining of alternate orders. We found that a useful mode of thinking this through, so as to measure political effects, was to scrutinize the relative (in)validity of the distinction between carnival and non-carnival time. Another was to look at the dialogic engagements and fragmentations of normative time and spatial demarcations within carnivals. A third tack was to occasionally laugh at our own verbosity.

Holistics: carnivals and their respective cultural systems

The first section, 'Carnivals and their respective cultural systems', presents three studies of prominent carnivals (Trinidad, New Orleans, Bahia) in the socio-anthropological line. Gerald Aching's 'Carnival time versus modern social life: a false distinction', suggests that carnival reveals complex intersections of gazes and performances of identity, in particular in the way women have used their bodies as instruments of protest and resistance. While *winin'* (dancing with 'lewd' pelvic gyrations) is associated by some with objectification of women and commodification of the carnival, for Aching this assertive, satirical and ludic dancing is consistent with an enduring, feministically-grounded struggle, and best read in concert with the complexity of modern social life.

Stephen C. Wehmeyer, in 'Feathered footsteps: mythologizing and ritualizing Black Indian processions in New Orleans' demonstrates the links between New Orleans Spiritual churches and Black Indian tribes. While these are usually perceived as belonging to separate spheres, the Black American religious institutions appropriate into their pantheon of spiritual guides the entity of Black Hawk, who is also embodied in carnival street performances of Indian tribes on the streets of New Orleans. These interdependent performances reflect the intrinsic relation between the secular and the profane, between the carnivalesque and everyday order. Both groups suffer external pressures as institutions, and attempt to generate a productive performative space for their members. This touches on the issue of the putative authenticity of particular cultural traditions.

In 'Bahian carnival and social carnivalesque in trans-Atlantic context', Piers Armstrong continues the inquiry into the performative relation between carnival and conventional order, in Bahia in Northeastern Brazil. He describes carnival there as a subjective and eccentric culture as well as an economic field, a catalyst for symbolic action which condenses cultural abstracts into concrete forms and praxes, and sees collusions between subalterns and elites despite their seemingly opposed agendas and social separation. Bahian carnival is presented as an intensification of everyday culture; in compensation, he suggests carnival motifs and values have penetrated everyday culture so that the two exist in symbiosis.

Hierarchies: carnival histories and state interventions

The essays by Philip Whalen and Kristen McCleary, both historians, focus respectively on France and on Argentina and Uruguay. They focus on how State elites appropriate and commandeer popular culture, and the connivance of the populace in those strategies. Whalen's 'The return of Crazy Mother: the cultural politics of carnival in 1930s Dijon' explores the innovative and devious strategies of the political elite in inter-war France to promote its modernization agenda by simultaneously nurturing festive traditions. The State incorporated carnival and carnivalesque motifs to industrial projects. Regional folkloric traditions served as palliatives to national narratives maligned in the First World War. In Whalen's account, carnival and carnivalesque are susceptible to fabricated narratives (i.e., convenient to external interests), and consequently a certain time/space unity of culture, politics and a variety of official and unofficial agendas is successfully forged. Whalen describes the carnival parades as heteroglossic moments which permit the emergence of unscripted and unchoreographed moments, and, finally, a middle space constituted, simultaneously by an uncertain domination and strategies of accommodation, symbolic inversion and aesthetic evasion.

Not all such manipulations go uncontested. Kristen McCleary's 'Ethnic identity and elite idyll: a comparison of carnival in Buenos Aires, Argentina and Montevideo, Uruguay, 1900–1920' highlights the differences in the evolution of carnival in Argentina and Uruguay, where the former's carnival traditions were stamped out, and the latter continued its performances despite official efforts at suppression. McCleary analyses the durability of carnival as contingent on its public (official and/or popular) reception. Perceived as a potentially disorderly and disruptive event, Argentina's suppression of carnival was enabled by the municipal structure of Buenos Aires, which had tremendous control over urban public spaces. In Montevideo, in contrast, due to a less stable political system with few such controls, carnival's practitioners held sway over performances. McCleary describes local and national debates for each case. While often neglected in more theoretically minded socio-anthropological accounts, such details are crucial in understanding the mechanics of carnival. To a surprising degree, the qualitative colour of virtually all carnivals is contingent on municipal organization.

Horizons: transplanted carnivals, and carnivalesque strategies in contemporary cultural politics

A useful way to test arguments about the general character of carnival which are developed by scholars embedded in particular carnivals is to see how well such arguments pertain in the carnivalesque margins of carnival, i.e., in non-traditional sites in which carnivalesque motifs are consciously invoked by participants or where their features warrant such invocation by scholars in the process of analysis. The last section presents an essay on a carnival transposed in the diaspora of the diaspora, and two more essays on carnivalesque (non-literal carnival) scenarios.

Lesley Ferris' essay, 'Incremental art: negotiating the route of London's Notting Hill carnival', deals with the Notting Hill Carnival in London – which is loosely based on Trinidad carnival, but is not held during the traditional pre-Lenten dates. She focuses on the production of carnival itself, affirming carnival's re-spatialization and general displacement of the everyday, its behavioural extraordinariness and its occupation of the street as its ultimate stage. Her analysis underscores the place of carnival in the arts and especially in theatre. Metropolitan ex-colonials lead this self-consciously multicultural spectacle, reframing identities and subjectivities and militating against the separation of the metropole and the former colonies and their respective histories. Focusing on a particular *mas* artist, Ferris transports her readers *en route*. She legitimizes carnival *mas* as street theatre and then explores the existential frame of street theatre, taking up the debates of Errol Hill and Derek Walcott as to whether carnival is theatre – the former seeing carnival as a 'mandate' for a national theatre, the latter as a theatricality which was not legitimate theatre (see Hill, 1972).

L.M. Bogad coins the phrase 'tactical carnival' to discuss the deployment of carnivalized protests in the global justice movement. This may remind the reader of the sorts of 'tactical interruptions' by lesbian and gay activists, in particular the members of ACT-UP (the AIDS Coalition to Unleash Power) during the 1980s to draw attention to the inadequacy of public funding for AIDS research. That in turn reminds us of the practice of 'outing'... at which point we have probably drifted too far from the humour and the point of the carnivalesque. The burlesque hero of Bogad's 'Carnivals against capital: radical clowning and the global justice movement' is CIRCA, the Clandestine Insurgent Rebel Clown Army, whose minions draw attention to global devastations and dislocations caused by neoliberal world trade policies (such as those protested more conventionally at WTO meetings). Bogad proposes that these protesters are 'attempting to... create a new, twenty-first century kind of "carnival" that is not calendrically nor spatially circumscribed or permitted by the state but declared and embodied by a movement that identifies itself as global, anti-corporate and anti-authoritarian'. They present a globalized identity in that they do not advance grievances specific to any one place or geography, but rather a singular plural identity.

The last essay in this collection is the closest in spirit to Bakhtin's characterization of the carnivalesque in the France of Rabelais. It is not about people who participate in a recognized carnival or self-consciously call their practices carnivalesque, but rather about a proletarian community which, the essayist argues, is effectively enacting the carnivalesque. Mary T. Hufford's 'Carnival time in the Kingdom of Coal' describes the struggles of between the ecological movement in West Virginia's coal districts and the industry bosses. The region is notorious both for poverty and lack of education, for brutal inroads on the landscape (including the levelling of mountains and the clogging of waterways) and for diverse forms of shocking chemical pollution. Hufford develops the metaphor of ol' King Cole (as malevolent industrial master) and the communitarian populace (as the good 'folk'), and traces the colourful and jesting protests of the latter. These have occurred over some decades. The longevity of this struggle and the intractability of the rapacious companies give it something of the old time flavour of lore. This same time frame of eternal conflict does not fit the usual cyclical culmination of carnival. Rather, Hufford's extended transposition of Bakhtinian motifs into what is overtly a harsh social struggle give us a rare opportunity to see how these fabled motifs might actually pertain in our own back yard.

Conclusion

All of the very different approaches of the contributors to this special issue read the relationship between the carnivalesque and the carnival as they do the social and the political, that is, as two sides of the same coin. Carnival traditions, along with other traditions, can be considered embodied arguments. But carnival may be different, as a restless yet timeless *geist*. Its playful demand for internal oppositions, for destabilization and improvisation, tends to precipitate mutation, though not abandonment, by its followers. Because it is predicated on established performance genres which often coincide with social or ethnic niches, performance tends to reiterate certain identities, and the participants involved experience an existential plenitude in which the free play of the ludic is balanced against a ritual confirmation of identity. And yet that difference is not easily incorporated into the whole either by the acquisition of a real hegemony (an aspiration which is constantly deferred as each group propels itself to centre-stage, achieving a momentary utopian hegemony), or by the opposite – assimilation to the mainstream. It is as if a centrifugal force in favour of difference stood counterpoised to the felicitous and promiscuous gregariousness of *the people*. Such is the magic of carnival – a profane ritual which affords the sensation of a rare state of singular social presence and anchors participants in an atavistic axis even as it portrays utopias.

Acknowledgements

This collection derives from a 10-day symposium, 'The Arts and Cultural Politics of Carnival', held at the Obermann Center for Advanced Studies at the University of Iowa in 2005. Michaeline A. Crichlow was the co-convener of the Iowa symposium. She expresses her thanks to Jay Semel, the Center's director, and his very capable staff who generously accommodated our needs; to Loyce Arthur of the Theatre Department at the University of Iowa, who co-directed the symposium; to Rob Sikorski, Director of the Duke University Center for International Studies (DUCIS), who underwrote the costs of the images used in this volume; to Tanya Shields, Judith Bettelheim, Fragano Ledgister and Patricia Northover, as well as to the anonymous readers of *Social Identities*, for their detailed and insightful comments on the various articles.

References

Aching, G. (2002). *Masking and power: Carnival and popular culture in the Caribbean*. Minneapolis: University of Minnesota Press.

Armstrong, P. (2001). Moralizing Dionysus and lubricating Apollo. A semantic topography of subject construction in Afro-Bahian carnival. *Luso-Brazilian Review*, *38*(2), 29–60.

Averill, G. (1994). Anraje to Angaje: Carnival politics and music in Haiti. *Ethnomusicology*, *38*(2), 217–247.

Bakhtin, M. (1968). *Rabelais and his world*. (H. Iswolsky, Trans.). Cambridge, MA: MIT Press. (Original work published 1965)

Bettelheim, J. (2001). Carnival in Santiago de Cuba. In J. Bettelheim & F. Ortiz (Eds), *Cuban festivals: A century of Afro-Cuban culture* (pp. 94–126). Kingston, Jamaica: Ian Randle Publishers; Princeton, NJ: Markus Wiener Publishers.

Bourdieu, P. (1990). *The logic of practice*. Stanford: Stanford University Press.

Burke, P. (1987). The carnival of Venice. *The historical anthropology of early modern Italy: Essays on perception and communication* (pp. 183–190). Cambridge: Cambridge University Press.

Butler, J. (1993). Critically queer. *A Journal of Lesbian and Gay Studies*, *1*(1), 17–32.

Clark, K., & Holquist, M. (1984). *Mikhail Bakhtin*. Cambridge, MA: Harvard University Press.

Crichlow, M.A., with Northover, P. (2009). *Globalization and the post-Creole imagination: Notes on fleeing the plantation*. Durham: Duke University Press.

DaMatta, R. (1991). *Carnivals, rogues, and heroes: An interpretation of the Brazilian dilemma*. (J. Drury, Trans.). Notre Dame, IN: University of Notre Dame Press.

Drewal, M.T. (2000). Nomadic cultural production in African diaspora. In N. Mirzoeff (Ed.), *Diaspora and visual culture: Representing Africans and Jews* (pp. 115–142). London & New York: Routledge.

Emerson, C. (1997). *The first hundred years of Mikhail Bakhtin*. Princeton: Princeton University Press.

Freyre, G. (1963). *The mansions and the shanties*. (H. de Onís, Trans.). New York: Knopf. (Original work published 1933)

Gilroy, P. (1993). *The black Atlantic. Modernity and double consciousness*. London: Verso.

Hill, E. (1972). *The Trinidad carnival; Mandate for a national theatre*. Austin: University of Texas Press.

Hirschkop, K. (1999). *Mikhail Bakhtin: An aesthetic for democracy*. Oxford: Oxford University Press.

Hirschkop, K., & Shepherd, D. (Eds). (2001) *Bakhtin and cultural theory*. (2nd ed.). Manchester: Manchester University Press.

Hitchcock, P. (1993). *Dialogics of the oppressed*. Minneapolis: University of Minnesota Press.

Irobi, E. (2007). What they came with: Carnival and the persistence of African performance aesthetics in the diaspora. *Journal of Black Studies*, *37*(6), 896–913.

Kerrigan, D., & Laughlin, N. (2004). The dramatist. *Caribbean Beat: The Magazine of the True Caribbean*. Retrieved June 10, 2008, from http://www.meppublishers.com/online/caribbean-beat/archive/index.php?pid = 6001&id = cb65-1-62

Largey, M. (2000). Politics on the pavement: Haitian Rara as a traditionalizing process. *The Journal of American Folklore*, *113*(449), 239–254.

LeRoy Ladurie, E. (1979). *Carnival in Romans*. (M. Feeney, Trans.). New York: G. Braziller.

Mbembe, A. (1992). Provisional notes on the postcolony. *Africa*, *62*(1), 3–37.

Mbembe, A. (2001). *On the postcolony*. Berkeley: University of California Press.

Real, K. (2004). Evoé! The carnaval of Recife and Olinda in Pernambuco, Brazil. In M. Barbara (Ed.), *Carnaval!* (pp. 203–237). Washington, Seattle: University of Washington Press.

Riggio, M.C. (Ed.). (2004). *Carnival: Culture in action: The Trinidad experience*. New York: Routledge.

Rohlehr, G. (2004). Calypso reinvents itself. In M.C. Riggio (Ed.), *Carnival: Culture in action: The Trinidad Experience* (pp. 213–227). New York: Routledge.

Ruiz, T.F. (2001). From carnival to Corpus Christi: Festivals of affirmation. In *Spanish society, 1400–1600* (pp. 143–162). New York: Longman.

Schechner, R., & Riggio, M.C. (1998). Peter Minshall: A voice to add to the song of the universe. *TDR–The Drama Review*, *42*(3), 170–193.

Scher, P.W. (2002). Preservation, carnival and the state in Trinidad. *Anthropological Quarterly*, *75*(3), 453–484.

Stallybrass, P., & White, A. (Eds). (1986). *The politics and poetics of transgression*. Ithaca: Cornell University Press.

Stam, R. (1989). *Subversive pleasures: Bakhtin, cultural criticism, and film*. Baltimore & London: Johns Hopkins University Press.

Todorov, T. (1984). *The conquest of America: The question of the other*. (R. Howard, Trans.). New York: Harper & Row. (Original work published 1982)

Turner, V. (1969). *The ritual process. Structure and anti-structure*. Chicago: Aldine.

Turner, V. (1987). Carnival, ritual and play in Rio de Janeiro. In A. Falassi (Ed.), *Time out of time: essays on the festival* (pp. 74–90). Albuquerque: University of New Mexico Press.

Walcott, D. (1970) What the twilight says: An overture. In *Dream on monkey mountain* (pp. 3–40). New York: Farrar.

Carnival time versus modern social life: a false distinction

Gerard Aching

Romance Studies, Cornell University, Ithaca, USA

Closely examining the dance form of *winin'*, ubiquitous in Trinidad and Tobago's carnival, this essay argues for the inextricability of carnival time and contemporary social life. In contrast to the notion that carnival constitutes interruptions or postponements of projects of modernity and, especially, that it invokes a temporality and social space where ideologies may be blissfully suspended, this study illustrates how this dance form articulates the status of and quest for personal freedoms in public spaces and contests a specific gender ideology. The essay describes and interrogates how *winin'* mediates the relationship between competing pleasures – those of the state and of the carnival reveller respectively – and illustrates the extent to which the dance form's exaggerated and hypervisible practices constitute a demand for social engagement.

The centrality of carnival to Caribbean cultures cannot be exaggerated. Antonio Benítez-Rojo posited this idea claiming that 'of all possible sociocultural practices, the carnival... is the one that best expresses the strategies that the people of the Caribbean have for speaking at once of themselves and their relation with the world, with history, with tradition, with nature, with God' (Benítez-Rojo, 1997). Masks and masking practices mediate such relationships, which should not be surprising because few objects so essential to carnivals everywhere have been more germane to areas of subjectivity and aesthetics. Indeed, with respect to most narrations of subjectivity, it would not be an exaggeration to claim that masks and their rhetorical manifestations have consistently been instrumental for representing forms of recognition, self-recognition, and misrecognition.

If carnival has proven to be a notoriously elusive object of critical inquiry, one of the main difficulties in arriving at a scholarly assessment of it resides in the consistency with which a particular theoretical conundrum has been attributed to this social phenomenon. On one hand, carnival has been defined as an event that frames, stages, and foregrounds marginalized and frequently subversive social spaces and identities; on the other, it is often assumed that this alterity is not only unsustainable beyond the event, but that its presence in public spaces ultimately contributes to social stability. At the heart of this conundrum, there lies a specific ambiguity: a mechanistic notion of carnival as a safety valve that contributes to maintaining social order through subversive but contained acts or, at least, gestures.

Most ontological approaches posit carnival as a perpetual simultaneity and imminence of binary oppositions or as a site of overdetermined conduct in public spaces; the majority of these approaches juxtapose carnival time to the routines of daily life. The foremost obstacle to theorizing carnival, I would argue, has been the broad acceptance that carnival time and modern social life are mutually exclusive and, by the same token, that carnival invokes a temporality and space in which ideologies may be blissfully suspended.

We impoverish our understanding of carnival by conceptualizing it in these limited ways. In arguing against these assumptions, I want to promote a shift in analysis away from the preponderant notion that carnival consists of simple high-low inversions and that its revelry in the Americas is reducible to a pre- or anti-modern vestige of colonialism that interrupts or postpones projects of modernity. Fundamentally, I want to propose a more complex understanding of the relationship between carnival practices and modern social life through a reflection on carnival time as a means of articulating the status of and quest for personal freedoms in public spaces. The context for this reflection is Trinidad and Tobago's annual carnival, which began over two hundred years ago and which has become inextricably tied to the country's national identity (see Hill, 1972; Cowley, 1996; van Koningsbruggen, 1997).

Some influential thinkers have argued against separating carnival from modern life. For example, C.L.R. James asserted that it was not accidental that the expansion of carnival in Trinidad coincided with the development of democracy and nationalism (James, 1992). At a historical juncture in which independence from Britain was imminent, James argued that local government should facilitate carnival because the event gave rise to spontaneous organizational skills that he associated with the potential for self-government. In another instance of cultural criticism that belonged to this pre-independence period, James suggested that future sociological and psychological research on carnival will ultimately view the energies that culminate in the event as a substitution by the people for a national activity removed from the social life, national festivals, and other national activities 'which were imposed on them from the outside' (James, 1992). The distinction that James made between carnival and habits imposed from outside emerged from a commonly-held view that colonialism pervaded and dominated all aspects of social life. James reiterated this idea when he stated that carnival's 'diversion of energies' from ordinary social work 'goes very deep into the history of islands like Trinidad and others in the West Indies which are westernised without any nationalisation of their own' (James, 1992). Even though James attempted to associate carnival with the emergence of a modern nation-state, his pre-independence argument that social life is imposed from the outside left little room for an analysis of complex, local subjectivities and pre-Independence nationalisms within the island's colonial society.

In 'Bakhtin, Marxism, and the Carnivalesque', Dominick LaCapra claims that carnival's social function relies on a perennially shifting and ambivalent reciprocity between carnival time and ordinary life (LaCapra, 1983). Asserting that carnival should be considered much more than a mere safety valve, LaCapra argues that 'the existential importance of festive carnival forms in certain cultures and groups might lead one to reverse standard functionalist preconceptions and to interpret "ordinary" life in terms of the way it prepares for carnival' (LaCapra, 1983). Indeed, the fact that carnival is the mainstay of tourism in Trinidad and that some of its costume designers are at the forefront of the nation's culture industries underscores carnival's influence outside its annual period of celebration and locates it fully within the country's modernization projects. Similarly, Juan Flores poses a question about popular culture in *From Bomba to Hip-Hop* that challenges the conceptual split between popular culture and modern-day society. Is it possible, Flores asks, to engage '"moments of freedom"... without positing some space outside of and unaffected by the industrial, ideological, and mobile demographic conditions that so obviously prevail in contemporary society on a world scale?' (Flores, 2000). Flores answers that these moments of freedom do not belong to an alternate space or time. What is particular about popular culture, he affirms, is that it is momentary, and even as its traditions derive from a historical past, 'it is present, it is contemporary, it is always *now*' (Flores, 2000).

If our conception of carnival is weakened by a notion of it as little more than a safety valve that suspends ideologies by giving way to festive abandon, then what kinds of critical approach would provide a deeper analysis of the relationship between carnival time and contemporary social life? In order to avoid the simplifying high-low paradigm that is frequently attributed to Mikhail Bakhtin's work on the carnivalesque, I would like to refrain from pitting the postcolonial nation-state against an undifferentiated citizenry (the 'people', the 'masses'). A useful approach would be to examine carnival time according to what may be called 'competing pleasures' – that is, pleasures through which the nation-state, defined in part by legislation that 'facilitates' and circumscribes carnival pleasures, is engaged in complex relationships with carnival revelers who perform their 'moments of freedom' in public spaces during the festivities. In defining 'productive pleasures' as moments that escape theorization because they are transitory, John Fiske revisits the distinction that Roland Barthes made in the 1970s between pleasure (*plaisir*) – that is, a mundane pleasure based on the recognition, confirmation, and negotiation of social identity – and bliss (*jouissance*), which Barthes identifies with the ecstatic evasion of the social order, the pleasure of the body at the moment when culture gives way to nature (Fiske, 1989). Fiske refers to bliss as a 'loss of self' and 'evasion of ideology' but not without critiquing the ideological limitations of Barthes' terms: he states, with respect to bliss, that '[t]he evasion is experienced as empowering, but does not determine, or necessarily even influence, the use to which that empowerment may be put' (Fiske, 1989). What is implied here is that, even though the individual carnival reveler might experience a momentarily ecstatic and empowering 'loss of self', this release is nonetheless vulnerable to and circumscribed by the political and ideological uses for which his or her bliss may be recruited.

With respect to Trinidad's carnival, the opinion that bliss is readily available to all revelers can be seen as an ideological position – that is, as a desired and promoted fiction that mystifies social relations under the guise, for example, of cultural and national unity. Earl Lovelace unmasks this fiction persuasively in *The Dragon Can't Dance* (1979). In this classic novel about carnival in Trinidad, the characters who joyously proclaim unity ('all o' we is one') are precisely those who, for mostly economic reasons, distance themselves from their community but return anxious to affirm that they still belong to it. Today, the desire for abandon during carnival, particularly in the form of provocative *winin'* and/or marathon dance performances, functions for many local and expatriate Trinidadians and Tobagonians as the expression and performance of an attitude toward a cultural and/or national identity.

Winin' (winding, gyrating, and thrusting the pelvis) is a provocative dance performance that is ubiquitous at carnival time and frequent at *fêtes* (parties) during the rest of the year. It was originally associated with the urban 'underclass' during the nineteenth and early twentieth centuries but is today far less marginalized because of the heavy participation of individuals from a broad range of social classes. This performed mode of identity is also the scrutinized object of technologies of (in)visibility and image manipulation that work through cultural tourism to sustain certain nationalist ideologies and discourses. Consequently, a worthwhile and flexible distinction can be made between the reveler's individually empowering but potentially vulnerable bliss and the state-regulated pleasure of inclusion within a cultural and/or national identity to which the reveler may (or may not) subscribe.

Viewing carnival time as instances of competing pleasures has a particular advantage. It permits us to go beyond a conceptually unwieldy antagonism in which the nation-state and the 'popular masses' are considered autonomous social actors at the same time that their engagements are stereotypically reduced to confrontations between a subjugating nation-state and the subjugated but resistant 'popular masses'. The notion of competing pleasures allows us to take into account diversely situated rival agents and agendas as well as varying degrees of personal and collective self-consciousness. In this essay, I would like to focus on the notion of bliss that is frequently associated with carnival and examine how this apparently ecstatic 'loss of self' invokes the status of and desire for personal freedoms in public spaces during carnival. Bliss, I want to emphasize, should not be equated with freedom itself. Rather, it represents a strong drive toward individual freedoms that can be performed, observed, and, because it appeals for a response, answered in the street. Bliss is the result of a psychic and bodily desire to transcend the gap between a present moment and the (ever elusive) experience of absolute freedom, and carnival's appeal lies precisely in its purported potential to invoke this time and place of absolute freedom. In local parlance, to 'free up' refers to a letting loose, to a relaxation of established 'middle-class' codes of personal conduct, especially with regard to the body, at the same time that the middle class has come to dominate public exhibitions of carnival practices (see MacDonald, 1986). *Winin'*, with its pelvic gyrations and thrusts, is perhaps the most visible example of this desire for a 'loss of a self' in public that may be perceived as subject to specific social codes and pressures.

In this essay, my fundamental argument is that carnival time represents not the absence but the exacerbation of social antagonisms. In making this claim, I want to suggest that what lies at the heart of the blissful abandon that I have described is a broader notion of freedom than the kind that is currently invoked to speak about carnival time. Unlike Bakhtin's concepts of the carnivalesque and the grotesque, which locate the agency of the popular masses in the ambivalence of its always excessive or unfinished bodies (Bakhtin, 1965/1984), the broader definition of freedom to which I refer describes a bodily self-mastery that externalizes itself as 'exaggeration', 'overdetermination', and 'hypervisibility'. My contention is that even though Bakhtin's writings on such aspects of popular culture have been useful for their sustained examination of the body, they provide little information about the carnival reveler's agency. If we are to approach the complexities of carnival time in the modern world, it is important to begin to imagine this reveler not as an exotic 'other' whose agency is reducible to an essentialized ambivalence that elides all forms of social coercion but as an agent who seeks the blissful experience of bodily self-mastery in spite of them.

Freedom and visibility

Before proceeding, it is necessary to be more concise about the notions and elements of freedom that I invoke in this study. My foremost assumption is that absolute freedom does not exist. As long as one lives in a society, there is no place within it that is either entirely free of coercion, no place where it is possible to be supreme master of the conditions of one's existence. Given this assumption, the question of freedom becomes one of distinguishing individual freedoms from the idea of overall freedom, which can be understood, not as synonymous with absolute freedom, but as an aggregate of specific freedoms. In both cases, the idea that one simply possesses or lacks freedom is a vague statement, because freedom is fundamentally a 'relational concept' – that is, its measurement emerges from a relation between a specific agent or subject who is free *vis-à-vis* another agent to perform a particular action (Carter, 2004). Statements, therefore, like 'I am free' or 'I am oppressed' do not provide sufficient information about a subject's freedom unless they are accompanied by an explanation of the relationship between the subject, a rival agent, and a specific action.

Another important aspect of freedom that I incorporate in this discussion is the subdivision of freedom into positive and negative notions. According to the political theorist, Isaiah Berlin, who discussed this subdivision in a seminal study, we typically make use of two concepts of liberty in the western tradition of liberalism. The first and more familiar concept seeks an answer for the question, in what areas of life can a subject act without interference from others? (Berlin, 2002). Berlin calls this concept the negative sense of liberty because it concentrates on describing *freedom from* other persons or from institutions. For Berlin, the positive concept of liberty would provide an answer to the question, who or what is the source of control that can determine a subject's actions? This concept is positive because it designates the *freedom to* be one's own master (freedom as self-mastery). Even though these two concepts of liberty are fundamentally concerned with the actions of maintaining coercive agents at bay, they are sometimes opposed. Freedom from coercion, for example, does not directly lead to self-mastery. More significantly, it is usually the case that securing freedoms for the greatest number of individuals in a community generally means compromising one's own (Berlin, 2002).

In the literary scholarship on carnival, the issue of individual freedoms is normally (but only implicitly) addressed by a focus on aesthetics. An almost *de rigueur* recourse to Bakhtin's *Rabelais and His World* is one of the principal reasons why critics may not have taken the scholarship on carnival beyond the Russian intellectual's aesthetic theories. Undoubtedly, his conceptualization of the carnivalesque and the grotesque was seminal for opening areas of research on the literatures and cultures of early modern Europe; but the description and critique of modernity that emerge in Bakhtin's study are not universally applicable to the diverse contexts for which they have been employed. Through no fault of his own, he posited the grotesque as troubled mediations between so-called high and low cultures in ways that were later recruited for some postcolonial analogies in which 'high' became synonymous with the colonizer's culture and 'low' with that of the colonized.

Bakhtin's categories have readily been employed to describe cultural and literary contexts that, by virtue of their distance from Europe, simply challenged established, metropolitan epistemologies. The ambivalence and elusiveness of the carnivalesque and the grotesque, in short, have too frequently been associated with the irreducible difference of exotic 'others'.

Yet Bakhtin's aesthetic concerns were intimately related to the question of individual freedoms in a very specific context. Michael Holquist argues that we should visualize grotesque realism as the Russian intellectual's meticulous attack on Soviet socialist realism in the 1930s (see his Prologue to Bakhtin, 1965/1984). Hence, what today might be considered theoretically unwieldy – that is, Bakhtin's identification of the grotesque with an overly optimistic populism – may be seen as a politically inspired overdetermination of the so-called masses, a purposefully exaggerated defense of the masses in the midst of a militarized cultural policy that included the control and censorship of literature and art (for a critique of this populism, see Stallybrass & White, 1986; LaCapra, 1983). The grotesque and the carnivalesque, in other words, are aesthetic categories that Bakhtin elaborated not only to interpret aspects of 'public culture' in early modern Europe but also to respond to the dictates of Soviet socialist realism. If we fail to keep this surreptitious use of the grotesque and carnivalesque in mind, we risk underestimating the ideological and political importance of 'exaggeration', 'overdetermination', and 'hypervisibility' for Bakhtin's aesthetic categories.

'Freeing up'

Let us now turn our attention to the significance of forms of 'excess' such as 'exaggeration', 'overdetermination', and 'hypervisibility' at carnival time in Trinidad and examine them from the vantage of those interrelated elements that inform us about situations of individual freedom: an agent (the *winin'* dancer), an action (*winin'* in public spaces), and a rival agent (established codes of 'proper' conduct in the public sphere). Even though not all women *wine*, nor are all *winers* women, the *winin'* female dancer is a ubiquitous figure at carnival time. I employ this figure here not to essentialize her as the blissful subject *par excellence* but, heuristically, as a guide through my argument. The sheer numbers of women who participate in the annual festival partly justify my selection. Men outnumbered women in carnival bands until the 1950s (van Koningsbruggen, 1997); since then, women have come to predominate in carnival bands in such a way that the topic of women's participation in the festivities so often turns into a local discussion about *winin'* that it is difficult to separate both issues (van Koningsbruggen, 1997). This coincidence of increased participation and the ubiquity of *winin'* invites an analysis of what appears to be a displacement of women from the 'respectability' of domestic space to exhibitionism in the street (see Wilson, 1973). Women have steadily been enjoying higher incomes, with the result that carnival bandleaders have tended to create costumes with these clients in mind ('pretty mas' and skimpy costumes locally referred to as 'dental floss') and, inadvertently, to marginalize men in the bands (van Koningsbruggen, 1997). Nevertheless, the debate about women *winin'* in the street at carnival time has gone beyond economic indicators and incorporates topics such as the transformation of public space, class, gender, and racial formations, sexuality, the crisis of male identity, and national definitions of morality.

Winin' is a form of dancing that has traditionally been associated with the *jamet* or *jamette* – that is, the urban 'underclass' and, especially, the figures of ill-repute who lived below the *diamètre* (diameter) of respectability in Trinidad's late nineteenth and early twentieth-century colonial society. Traces of this historical association with the *jamet* can be found in today's frequent call by soca performers for revelers to go through the exaggerated motions of 'misbehavin', 'gettin' on bad', and 'mashin' up de place'. These activities might have sufficed to threaten the peace several decades ago, but today they form part of a call and response pattern between singers and dancing revelers. Almost always performed with carnival music of one kind or another, the gyrations and rapid pelvic thrusts that characterize *winin'* have generally been considered vulgar: they are visually provocative, clearly erotic in significance, and, because they are executed in the street, rather than in the privacy of the bedroom, they may be considered – depending on the observer's attitude – anything from tantalizing to grotesque. For a woman, *winin'* may be performed with another or several revelers of the same or opposite sex; and because it is normally understood that *winin'* in these circumstances parodies sex acts, a typical public response is laughter. In Trinidad and Tobago, it would be extremely daring, if not dangerous, for two men to *wine* together in public; two women doing so are generally not greeted with the revulsion that two men *winin'* together would be. Clearly, it is not enough to indicate that *winin'* exaggerates the body in universally homogeneous ways regardless of who *wines*; unlike Bakhtin's homogeneous aesthetic categories, other meanings of *winin'* accrue according to factors such as the *winer*'s gender, sexuality, neighborhood, and racial and class identities.

An area of inquiry that remains unexplored is the way in which *winin'* has shifted from being identified with the marginalized *jamet* to becoming a popular form of dancing for mostly young women revelers from a wide variety of social and racial backgrounds. This shift bears some relation not only to cultural politics just before and after the country gained independence from Great Britain in 1962 but principally, since then, to carnival's current status as the island's most popular national festival. In a longer inquiry, it would be necessary to interrogate the ways in which newly independent nation-states consolidate themselves socially by assimilating their most subversive cultural elements and transforming them into 'privileged' symbols of nationhood through sophisticated technologies of visibility. Peter Stallybrass and Allon White offered a glimpse of this process when they asserted that the British bourgeois rejection of the grotesque suggested that what was '*socially* peripheral' to that class was in fact '*symbolically* central' to its formation (Stallybrass & White, 1986). The prevalence of *winin'* at carnival time implies one or more of the following: that there has been a dramatic transformation in what might be considered socially peripheral in the country's contemporary life since political independence; that the socially peripheral and symbolically central necessarily (and naturally) coincide in occupying public spaces and attention once carnival assumed the status of the preeminent festival and visual representation of national identity; and/or that the socially peripheral has been purposefully recruited and made symbolic in order to carry out agendas that directly affect the control and uses of public space – which brings to mind Fiske's observation that even though bliss might be experienced as empowering, it is powerless to influence how it may be incorporated and put to subsequent use.

The fact that *winin'* has moved to center stage, as it were, should alert us to struggles over and changes in the definitions, control, access and, specifically, the commercial and democratic uses of public space. Even though it has traditionally been the case that purchasing a carnival costume is one way of participating in carnival, the commercialization of carnival bands and the increasing costs of purchasing costumes have reduced participants to those who have this expendable income. This commercialization has grown, so that even though inexpensive costumes are still available, it is also possible to purchase some of them from abroad (included within a carnival experience package) using internationally recognized credit cards. This increasing privatization of the festival has had certain consequences for those revelers who can no longer afford to participate in carnival: they become spectators of the national festival and/or they look elsewhere for alternate sites for festivities and self-expression. Controversies about public ordinances, the construction of infrastructure on public lands for carnival celebrations, and the uses of public spaces at carnival time are commonplace. Yet, if it is certain that the socially peripheral has successfully been co-opted by nationalist agendas at carnival time and during the rest of the year, then we are ultimately left with the troubling issue of what social spaces are left for dissent and subversion.

For many observers, it is the image of the woman who blissfully *wines* 'alone' – that is, without an obvious *winin'* partner – that has provoked a great deal of intrigue in the press and even among researchers. Commentators have been hard pressed to explain this abandon and its frequency regardless of the *winer*'s racial and class allegiances. In viewing *winin'* as an expression of women's sexuality, the anthropologist Daniel Miller concluded that even though *winin'* looks erotic, what gets enacted is basically 'a sexuality which does not require men; it is not lesbianism but autosexuality' (Miller, 1997). Miller's methods in arriving at this term were highly unusual: some of his observations came from social gatherings that he admits were purposefully staged for his benefit by his informants. He also limits his survey of *winin'* to low-income women even though he states that women from other classes were heavily involved. Defining 'autosexuality', Miller states that it is 'an expression of a free sexuality which has no object but itself, and most especially it is a sexuality not dependent upon men' (Miller, 1997). Van Koningsbruggen takes Miller to task on his analysis proposing that 'autosexuality' might be the result of a conundrum in which the latter fails to clarify whether women *wine* in order to exclude men or whether they do so because relatively few men join the carnival bands in the first place. Arguing that women do not reject sexuality by *winin'*, van Koningsbruggen claims that they are principally involved in parodying it and poking fun at phalocentrism (van Koningsbruggen, 1997). Although both researchers associate *winin'* with expressions of sexuality, they do not speculate what their conclusions might imply for local definitions of sexuality, gender politics, and contemporary debates about morality.

A reveler who *wines* in the streets at carnival time does so with and against a complex set of agents and agencies. Those who consider *winin'* excessive or vulgar at all times assume, from their perspective, a morally superior position. Even at its most tangibly coercive, this position does not emerge directly from legal sources but from an array of agents and agencies that include the tone of legislative tendencies, religious teachings, and behavior codes enforced in schools and homes. It would oversimplify matters to treat these sources as collectively constitutive of a homogeneous, middle-class 'establishment'. For not only is it difficult to define the 'middle class' in so hybridized a social context, these individual agents can themselves be duplicitous. For example, while one branch of the government presents and debates sexual offences ordinances, the National Carnival Commission facilitates products of cultural tourism in which women revelers are prominently featured. The country's culture industries promote carnival festivities locally and abroad, but they carefully screen the images that they employ to advertise the national festival. Warnings against carnival excesses may be heard from the pulpits of Catholic churches, but it was the Church that historically sanctioned carnival activities as a pre-Lenten celebration, especially in the 1880s when British attempts to do away with the French Creole celebration set off violent rioting (Anthony, 1989). Any remonstrance against a woman blissfully *winin'* 'alone' in the street would typically emerge from a perennially shifting configuration of these agents and others in which degrees of power and coercion can make themselves felt. All of the complexities that I have been describing in this section of my essay are fully in play at carnival time, so that the feeling of abandon that a reveler enjoys is unavoidably embedded in a social milieu that is thoroughly suffused with competing political and ideological agendas. On what basis, therefore, can a woman *winin'* alone achieve blissful abandon in this context?

In order to begin to answer this question, it is crucial to observe that in addition to the coincidence of the socially peripheral and the symbolically central on the national stage, what has also occurred is the feminization of *winin'* in public. As I indicated earlier, the fact that carnival costume designers have been catering to a mostly female clientele already provides an economic basis for the prevalence of women in carnival bands. This situation does not mean that the majority of the men disappear from the streets at carnival time, but that they have mostly become spectators and/or occasional revelers. Men still dominate in other areas of carnival, such as in steelbands and, less so, in calypso and soca performances and competitions. In her research on dancehall in Jamaica, Carolyn Cooper argues that some women have feminized 'slackness' (the eroticism of dancehall performances) through 'an innocently transgressive celebration of freedom from sin and law. Liberated from the repressive respectability of a conservative gender ideology of female property and propriety, these women lay claim to the control of their own bodies' (Cooper, 1995). As in dancehall 'slackness', the images of women *winin'* alone do not signify that these self-liberating impulses are directly responsible for a crisis of male identity. Apart from suggesting that there can be nothing liberating for men about these impulses, this argument fails to take into account that whether or not men *wine* in the street at carnival time, men and women share public space in ways that allow for a diverse repertoire of reciprocal play, provocations, and responses.

Cooper's argument above outlines statements about freedom that are worth pursuing at this point. Sin, law, and a conservative gender ideology are certainly opposed to 'slackness' and *winin'*; yet there are two elements in these statements that deserve reflection. First, the idea that 'slackness' can be an 'innocently transgressive celebration' suggests an agency that is still unaware of the consequences of its actions, a celebration that is meekly offensive, or both. It may well be that the reveler's desire for a 'loss of self' aims at a state of 'innocence' that would be 'transgressive' precisely because it dares, inadvertently or not, to disavow the politics and ideologies that envelop it at carnival time and during the rest of the year. Second,

Cooper cites two notions of freedom that Berlin also outlined in his essay: one is the freedom from the coercion of laws, codes of propriety, and so on (Berlin's 'negative' freedom) and the other entails laying claim to one's body (the idea of self-mastery that informs his concept of 'positive' liberty). As I mentioned at the start, Berlin asserts that there are connections and disjunctions between both forms of freedom: freedom from coercion does not necessarily lead to self-mastery and, because absolute self-mastery is impossible in any society, it is typically the case that we agree to assume degrees of coercion for the sake of the common good. Nevertheless, the relationship between both concepts of freedom should probably be thought of, not as binary opposites, but as a continuum along which the distance between freedom from coercion and self-mastery is constantly being negotiated and shifted. Such a proposition allows me to claim, like Cooper does with respect to 'slackness', that *winin'* may be seen as a drive toward *both* concepts of freedom since they constitute the same continuum. As such, it becomes possible to argue that blissful experiences like the abandon, 'freeing up', and 'loss of self' to which *winin'* accedes are not limited to *freedom from* all forms of coercion, including politics and ideologies. In the final analysis, the psychological and intellectual reduction of freedom to its negative valence alone gives rise to the false impression that carnival time is only about freedom from coercion; and it is this false impression that makes it possible to imagine that carnival time and modern life are mutually exclusive.

Looking again

When *winin'* is performed in its highest and wildest degree of abandon, the body is not simply relaxed or 'freed up' because the reveler is intensely focused on creating the widest possible range of smooth gyrations and forceful pelvic thrusts, which may be slow or rapid and typically in dialogue with music. That the *winer*'s bodily self-mastery may generally be considered 'exaggerated' or 'excessive' would come as no surprise – but 'exaggerated' and 'excessive' in relation to what? In a sense, everyday language fails us when it attempts to capture the ways in which 'exaggerations' – such as 'serious' *winin'* and a host of other openly subversive activities in public spaces – are not 'excessive' from the perspective of those who carry them out but entirely appropriate for the social visibility that they wish to acquire. According to Berlin, 'the desire for recognition is a desire for something different: for union, closer understanding, integration of interests, a life of common dependence and common sacrifice' (Berlin, 2002). If this is true, then revelers who *wine* 'alone' are not interested in isolating themselves from others but, rather, in seeking ways of engaging with and belonging to a society in which the socioeconomic differences between the peripheral and the symbolic are palpable and prone to manipulation.

Adjectives such as those that I have cited reveal the positions from and attitudes with which we observe bodies. Describing carnival activities as local spectators and/or academically trained observers invariably and unavoidably entails risks of objectification. Yet there are ways to be mindful that the objectification of the carnival revelers' bodies need not be the final result of analysis. For Kaja Silverman:

> There is nothing we can consciously do to prevent certain projections from occurring over and over again, in an almost mechanical manner, when we look at certain racially, sexually, and economically marked bodies. That does not, in and of itself, signify the

> failure of the ethical. The ethical becomes operative not at the moment when unconscious desires and phobias assume possession of our look, but in a subsequent manner, when we take stock of what we have just 'seen', and attempt – with an inevitable limited self-knowledge – to look again, differently. (Silverman, 1996)

The *winin'* body at carnival time deserves this second or different look. The purpose of this look would not be to re-aestheticize this body and its performance or provide it with its own special category. Rather, the presence of this body should act as a reminder that even in the most democratic public spaces, there will always be subjects or agents who desire social recognition because they want to belong to and participate in their societies in more satisfactory ways. A second look might engage with the reveler in a different way by acknowledging his or her invitation to secure and retain the freedom and right to be socially visible, regardless of income, at carnival time and during the rest of the year. In this light, carnival time does not lead us away from politics and ideologies: it brings them to the fore in direct and dynamic appeals for social engagement.

References

Anthony, M. (1989). *Parade of the carnivals of Trinidad, 1839–1989.* Port of Spain: Circle Press.

Bakhtin, M. (1984). *Rabelais and his world.* (H. Iswolsky, Trans.). Bloomington, IN: Indiana University Press. (Original work published 1965)

Benítez-Rojo, A. (1997). Caribbean culture: A carnivalesque approach. In A.J. Arnold (Ed.), *History of the literature in the Caribbean* (pp. 203–213). Amsterdam: John Benjamins.

Berlin, I. (2002). *Liberty.* Oxford: Oxford University Press.

Carter, I. (2004). *A measure for freedom.* Oxford: Oxford University Press.

Cooper, C. (1995). *Noises in the blood: Orality, gender, and the 'vulgar' body of Jamaican popular culture.* Durham, NC: Duke University Press.

Cowley, J. (1996). *Carnival, canboulay, and calypso: Caribbean traditions in the making.* Cambridge: Cambridge University Press.

Fiske, J. (1989). *Understanding popular culture.* London: Routledge.

Flores, J. (2000). *From bomba to hip-hop: Puerto Rican culture and latino identity.* New York: Columbia University Press.

Hill, E. (1972). *The Trinidad carnival: Mandate for a national theatre.* Austin: University of Texas Press.

Holquist, M. (1984/1965). Prologue. In M. Bakhtin, *Rabelais and his world* (pp. xiii–xxiii). (H. Iswolsky, Trans.). Bloomington, IN: Indiana University Press.

James, C.L.R. (1992). Lincoln, carnival, George Padmore: Writings from *The Nation.* In A. Grimshaw (Ed.), *The C.L.R. James reader* (pp. 281–295). London: Blackwell.

LaCapra, D. (1983). Bakhtin, Marxism, and the carnivalesque. In *Rethinking intellectual history: Texts, contexts, language* (pp. 291–324). Ithaca, New York: Cornell University Press.

Lovelace, E. (1979). *The dragon can't dance.* Burnt Mill, UK: Longman.

MacDonald, S.B. (1986). *Trinidad and Tobago: Democracy and development in the Caribbean.* New York: Praeger Scientific.

Miller, D. (1997). *Modernity: An ethnographic approach: Dualism and mass consumption in Trinidad.* Oxford: Berg.

Silverman, K. (1996). *The threshold of the visible world.* New York: Routledge.

Stallybrass, P., & White, A. (1986). *The politics and poetics of transgression.* Ithaca, New York: Cornell University Press.

Van Koningsbruggen, P. (1997). *Trinidad carnival: A quest for national identity.* London: Macmillan Education.

Wilson, P.J. (1973). *Crab antics: The social anthropology of English-speaking negro societies of the Caribbean.* New Haven: Yale University Press.

Feathered footsteps: mythologizing and ritualizing Black Indian processions in New Orleans

Stephen C. Wehmeyer

Core Division, Champlain College, Burlington VT, USA

This essay explores intersections between the ritual veneration of stereotypified Native American spirit figures by American Spiritualists and the performative traditions of 'Masking Indian' practised by urban African Americans in New Orleans' Mardi Gras processions. The author examines the ways in which communities in New Orleans, of both sacred and secular identity, employ images of Native Americans as icons of spiritual power and presence. This essay suggests some ways in which this process constitutes an instance of co-narration, in which clergy and congregants of Spiritualist/Spiritual churches – whose narratives of the Indian spirits find expression through interlacing oral and ritual performances – have helped to establish a sacred dimension for Indian processions in New Orleans, adding an overtly spiritual note to otherwise secular 'rites of territory repossessed'. Through an examination of the community response to the death of Big Chief Allison 'Tootie' Montana, and the first post-Katrina Mardi Gras in 2006, the article explores the ways in which Indian icons and imagery still stand for many New Orleanians as powerful signs of something in the soul that, to paraphrase a popular Mardi Gras Indian song, won't kneel and won't bow down.

Masking Indian and Indian spirits in New Orleans[1]

> The working class blacks who create the Mardi Gras Indian tribes collectively author an important narrative about their own past, present, and future... But their collective narrative goes beyond literature and folklore. It draws upon a myriad of contradictory images and icons to fashion a syncretic unity. In the aesthetics of their performance, the Mardi Gras Indians balance the competing claims of commercial culture and folk culture, of America and Africa, of resistance and accommodation, and of spontaneity and calculation. Their art stems less from ancient storytelling traditions or aesthetic intentions than from the necessity imposed upon them by oppressive social conditions. Their utopian projections originate less from abstract images of an ideal future than from a determination to read the lessons of solidarity and struggle from the past and present into the future. If the practices of story-telling and the qualities of criticism and creativity embedded within them are to survive in the modern world, it will likely be through the multi-dimensional practices of artists and individuals like the Mardi Gras Indians. (Lipsitz, 1988, pp. 99–121)

> The god of the American city – and this was the singular deity of a decidedly monotheistic faith – was order, and the rituals of this faith were the control and supervision of the streets. In the sight of this deity (or its representatives), almost everything that happened on the streets in immigrant and migrant communities was profane. The struggle for the streets was a fundamentally religious contest. (Orsi, 1999, p. 37)

In recent decades, folk outside of New Orleans have been made increasingly aware of the city's 'Black Indian' or 'Mardi Gras Indian' traditions. The increased awareness of what Lipsitz would term the Indians' 'collective narrative' has been primarily due to the efforts of the Indians themselves, telling their stories through visual, musical, and performative arts, collaborating with ethnologists and historians, photojournalists, filmmakers, promoters and producers (Berry, 1986, 1995; Blank, 1978; Jacobs & Kaslow, 1991; Martinez & Hinton, 1977; Smith, 1994). Museum exhibits, audio recordings, documentary films, and live performances of music and dance serve now as the vehicles for telling elaborately interwoven stories of *marronage*, race violence, cultural survival, mutual aid, and family history. Historians inside and outside these organizations continue to explore and untangle the web-work of relationships that link the Black Indians to other cultural institutions, in New Orleans and out of it. Much of the scholarship on the Indians addresses the perceived oppositional relationship to 'official' carnival processions, or treats the tribes' function as moral exemplars, as foci for community cohesion, as retainers of cultural memory. Early popular histories of the Indians told by observers and participants alike center around themes of resistance and the 'creolized interculture' of the Maroon encampments, crystallizing in romanticized visions of antebellum Afro-Amerindian relations (Martinez & Hinton, 1977). Later narratives focus on ties between Indian traditions and old-world African or Afro-Caribbean antecedents (Nunley & Bettelheim, 1988; Thompson, 1993; Roach, 1992).

A growing body of recent scholarship highlights both the subtle and overt relationships that exist between the Indian tribes and the Spiritual church: another institution known for its veneration of Indian spirits, which also occupies a well documented, but often ambiguous position in the religious and cultural landscape of present day New Orleans (Berry, 1995; Jacobs & Kaslow, 1991; Smith, 1992, 1994; Wehmeyer, 2000, 2007). Claude Jacobs, Andrew Kaslow and Michael Smith have documented the presence of Black Indian costumes in Spiritual church services, and Jason Berry has explored the links between the churches and the tribes through the common ground of jazz culture. Although oral histories mark the Indian tribes as an older institution in New Orleans, predating the formal founding of the Spiritual church by some forty years, the complex relationships between these two institutions should not be construed as identifying clear vectors of *origin*. As I have attested elsewhere, Anglo- and African-American Spiritualist practice has had a documented presence in the spiritual landscape of the Gulf South since at least 1853, and its vibrant traditions of honoring and entertaining Indian spirits have clearly influenced both the practices of the Indian tribes and the present day Spiritual church (Wehmeyer, 2001). Neither institution can be said definitively to have birthed the traditions of the other. Rather, these communities emerge out of a common complex of popular belief, ritual praxis, and pageantry, and continue to exist in a relationship of dynamic mutual interchange, where, to use Lipsitz's construction, each creates communal narratives concerning the Indian spirits which can in turn be incorporated into the histories of its counterpart.

Joseph Roach sees the march of the Indian tribes, in part, as an articulation of themes of 'frontier space – and its control by nomads' in which participants willfully occupy a space of 'constantly shifting borderlands', reenacting what Roach calls a 'rite of territory repossessed, not to assert permanent ownership, but temporary use' (1992, p. 476). If, as Robert Orsi suggests, the struggle for those borderlands, for the streets separating neighborhood from neighborhood, uptown from downtown, is a 'fundamentally religious contest', then it follows that the narratives would emerge from, or eventually develop, a sacral character. In subsequent pages, I wish to offer further exploration of what might be considered a process of co-narration, suggesting some ways in which Spiritual church members – whose narratives of the Indian spirits find expression through interlacing oral or ritual performances – have helped to establish a sacred dimension for Indian processions in New Orleans, adding an overtly spiritual note to the 'rites of territory repossessed'.

Although the tribes and the churches seem on the surface to be organized for very different purposes, one a community of performance, the other a community of worship, photo-ethnologist Michael P. Smith stresses their similarities at deep structural levels. These similarities are marked enough that Smith refers to Indian Practices, the didactic and celebratory weekly meetings attended by members of a given tribe as 'Another Kind of Church' (M.P. Smith, 1994, p. 89). Smith, Berry and Roach all quote more than one Big Chief describing Indian performance in the language of spirit possession, sacred rapture, and divine inspiration associated with the Spiritual church, inviting a closer look at this less visible, but equally vital counterpart to the Indian tribes.

The churches, like the tribes, are a primarily urban phenomena, and serve an almost exclusively African-American constituency. Their clergy blend Roman Catholic iconography and material culture (altars, statuary, priestly vestments, etc.) with energetic Evangelical preaching style and scriptural exegesis. Prophecy, spirit communion, 'laying on of hands', and other ecstatic forms of worship are employed by some of the most gifted ministers and bishops. To this blend is added a proliferation of tools and techniques for physical and spiritual healing drawn from several corners of the African diaspora, or from European mystical systems. A rich pharmacopeia of herbs, sacred oils, incenses and other *materia sacra* are incorporated into both the communal and private rites of the Spiritual church.

Since their heyday in the 1920s and 1930s, these churches have, to some degree, always occupied a marginal position in the New Orleans religious milieu. Like the communities it serves, the Spiritual church has managed to persist despite, perhaps even *because of,* its perceived marginality. Although most regular members of the contemporary Spiritual churches are African-American, numerous clergy-members serve as spiritual advisors to an ethnically diverse clientele. Bishop Oliver Coleman, my chief consultant in the Spiritual church, claims a substantial client base in the Midwest that is 'mostly white folks' (O.C. Coleman, personal communication. 10 December 1997). According to the oral and documented histories of the early days of the movement in New Orleans, Italian-Americans formed a notable presence in early church congregations and clergy. The small size of the Spiritual church community as a whole has helped generate a sense of interrelation between the congregants and clergy of individual churches. Bishops and ministers are itinerant as a rule and nearly always attend or assist at services at churches other than those over which they preside. This situation as a whole has helped generate a sense of inter-relation between the congregants and clergy of individual churches. On one particular Sunday, for example, I accompanied Bishop Oliver Coleman to the morning service at Israelite Divine Spiritual Church, after which we attended an afternoon communion service led by Bishop Victoria McSwain at Guiding Light Spiritual Church, and then made a congenial late afternoon visit to St. Daniel's Spiritual Church. While some degree of competition or rivalry between individual churches or clergy-folk may occasionally be apparent, the general good fellowship among church members and clergy suggests an amorphousness to this community, which will coalesce at different times and sites of worship, with 'Bishop X' presiding over her own services before enthusiastically joining the congregation at the church of 'Bishop Y'. Interestingly enough, M.P. Smith (1994) identifies this 'networking ritual' as virtually identical in style and function to that practiced by the Indian tribes during non-festival season, when one tribe will 'receive' another at its evening practices.

According to the prevailing story of the origin of the Spiritual church in New Orleans, the church was founded by Mother Alethea 'Leafy' Anderson, a holy woman, originally from Raceland, Louisiana, who relocated to Chicago and became a practising Spiritualist. Mother Anderson was led by the spirit to leave her new home and spiritual mentors in Chicago and return south sometime between 1918–1920 to establish a church in New Orleans. The stories about Mother Leafy's Spiritual victories are numerous (and often mirror those of the city's other legendary female spiritual leader, Marie Laveau). Battling against financial hardship, racial and sexual prejudice, Mother Anderson consecrated the first of the Spiritual churches (Eternal Life Christian Spiritualist Church # 12 on Amelia Street), and ordained the women who would follow her as church matriarchs. According to church oral history, it was Mother Anderson who introduced the Indian spirit to the churches . . . embodied in her personal spirit guide, Black Hawk. While it is likely that Anderson is responsible for establishing the veneration of this specific spirit within the syncretic ritual framework of the Spiritual church, long before the founding of ELCSC #12 in 1920, Indian spirits were staple elements in Spiritualist practice throughout the South, and Indian imagery and iconography were employed by African Americans and whites alike in the practice of myriad subaltern traditions of healing and folk magic.

The historical Black Hawk was a Sauk Fox war hero who sided with the British during the war of 1812, and later led a resistance campaign against the US government and military that ended with what has become known as the Black Hawk War in 1832. By 1833, the vanquished Black Hawk had completely captured the public imagination by upstaging President Andrew Jackson himself at a number of public appearances where he was presented as a 'hostage' after the recent war. In 1834, at the age of 66, Black Hawk dictated his autobiography to a government interpreter. He moved with his family to a settlement on the Des Moines river where he died in 1838.

Within 20 years of his death we find a number of spirits called Black Hawk making regular appearances as spirit guides and messengers at séances and Spiritualist demonstrations throughout the North. In this initially phase of Black Hawk's emergence as part of the Spiritualist pantheon, there exists little evidence, beyond the name, that these figures had any connection in the minds of their devotees with the deceased Sauk Fox war chief. Adopted by Mother Anderson as a personal tutelary, Black Hawk continues to be one of the most venerated saints in the contemporary Spiritual church, and his altars are among the most dramatic and beautiful examples of sacred assemblage created by church folk. He has since been joined by a constellation of other Indian spirits (Sitting Bull, Uncle Bucket, Yellow Jacket) some of whom are honored in many of the churches, and some of whom are idiosyncratic characters in the lives of individual Ministers.

The very facets of the church which offer divine comfort and sustenance to its members have been the source of considerable censure and misunderstanding from those outside the churches, both laity and clergy of other denominations. The strong sacramental elements in church worship, the proliferation of saints' statues, candles and icons, the use of blessed oils and incenses, as well as the powerful presence of women in the church hierarchy, leads many non-members to identify the churches with 'Voodoo', 'Hoodoo', or 'Conjure-work'. Whatever the scholarship of recent years has done to lift the prejudices with which these latter systems of belief are usually characterized, the terms still bear the stamp of negativity, or at best moral ambivalence, in the eyes of most New Orleanians. (More than one non-church member to whom I spoke in New Orleans shudderingly referred to the Spiritual churches as 'Voodoo churches'.) It is on the moral level (rather than the practical one) that Spiritual church members consistently (and firmly) draw a line between their practices of spiritual 'Work' and the potentially negative practices of Hoodoo or Conjure. Most church members are adept at quoting scriptural passages which support their sanctified use of paraphernalia like oils and candles, in defense of accusations of 'Voodoo' leveled by those of other denominations. Mother Fannie Magee, quoted in Claude Jacobs and Andrew Kaslow's *Spiritual Churches of New Orleans* (1991) asserts the Biblical underpinning for the use of candles in church ritual:

> I was reading the eighteenth chapter of Psalms, and around the twenty-eighth verse, I began to read: 'If you light a candle'. Not a light, or a lamp, but a C-A-N-D-L-E! A candle! Which so many people talk about and say 'Oh this is Voodoo'. But, if its Voodoo, it's what God ordained! That's why I'm happy over it! Because God ordained it, you know! Anything God does, I'm happy over it! And he said if you light a candle, you're going to light your world! (p. 56)

The presence of the Indian spirits in church ritual and iconography is often the focus of ambivalent attitudes from outside the church. Perceptions of Black Hawk as a potentially dangerous spirit, as a 'Hoodoo saint', help create an atmosphere of defensiveness around this figure. In one notable sermon I heard preached at the Black Hawk feast at Israelite Divine in 1997, Archbishop Johnson identified the multi-ethnic pantheon of the modern Spiritual church with the Pentecostal outpouring of the early Christian church:

> People ask us why we got an Indian in our church. Well, children, you should remember that when Jesus left his people for the second time . . . He told his apostles he was going away, but he would send them his divine spirit which was the *Holy Ghost*. And when he sent that spirit they began to preach and heal and speak in tongues . . . and there was all kinds of people in that room . . . there was Dagoes there, and Chinese, Japanese and all kinds of people. And everybody received the spirit. So its not so unusual that we should have an Indian guide . . . an Indian spirit in our church. (Archbishop E.J. Johnson, sermon, 10 December 1997)

Nonetheless, when testifying during services for the Indian spirits, one frequently hears such oblique statements as 'I don't call on him [Black Hawk] much ... But I know that those who call on him in faith will see justice!' Or similarly, 'I don't work that closely with Black Hawk, but those that do know that God will back up the spirit and work through him!' With the presence of elaborate altars for Black Hawk and the other Indian spirits in the churches, it is clear that *somebody* is working closely with Black Hawk. The admission of this fact at the personal level is another matter entirely, further testifying to the atmosphere of ambiguity and ambivalence which surrounds the material manifestations of the Indian spirit.

In many ways, this ambivalence is echoed the legacy of violence recalled by members of the Indian tribes.

> Violence at the hands of whites provoked in some Indians a militant strain, suppressed through normal life yet freed in Carnival. In the 1850s, Irish Channel whites openly terrorized blacks entering the area. By the 1920s, in the same neighborhood, whites dressed as cavemen fought the Indians with spike-studded cypress clubs. (Berry, 1995, p. 107)

Actual warfare between rival tribes, while heartily disapproved of in the present day, is part of the Indian cultural memory, recalled in narratives, ritual banter, and song lyrics: 'A long time ago we used to crack some heads, now we knock 'em dead with a needle and thread...' (Dollis, 2002). It is in part because of this checkered heritage, that the cry 'Indians Coming!' still has the potential to send 'chills down the backs of most downtown people' (Smith, 1994).

Civil rites: Indian processions and the ritual re/possession of urban space

> The legendary crisis in Mother Anderson's ministry came when she was arrested, according to many, for leading a band of marchers, barefoot and carrying candles, through the streets of the city without a permit... When brought before the judge, she prophesied that his child at home was gravely sick. After finding this warning to be correct, he dismissed the charges and, out of gratitude, helped her to obtain a state ministerial license. (Estes, 1998, p. 170)

Jason Berry (1995) offers another variant of this same tale, told to him by Bishop Edmonia Caldwell, Mother Leafy Anderson's niece:

> 'She [Leafy Anderson] had a candle drill', says Bishop Caldwell, 'She had a march, people were barefoot, and the police didn't allow you to have no marches at that time in the city'.
>
> Arrested and incarcerated, Anderson was eventually brought to trial, at which point, Bishop Caldwell maintains:
>
> She told the Judge, 'you should get to your house, your baby is sick'. The Judge's wife called and said the baby is sick. They rushed the baby to the hospital. Then she [Anderson] went back and her and the judge *talked*'. (p. 113)

During the course of my own field research, I have been repeatedly told versions of this story by a number of Spiritual folks in New Orleans, including Bishop Caldwell. All of my informants attribute this successful prophecy to Anderson's primary guide, Black Hawk, and nearly all mention that as a result the judge helped

Anderson obtain either the necessary credentials to operate a licensed church, or a permit to hold 'candle drills' and similar processions.

David Estes further comments that he has been unsuccessful in attempting to locate official records of Anderson's arrest and trial, although records do exist of another arrest, in April of 1925, at which police broke up a 'voodoo meeting' held at a private home, and presided over by one 'Lethe Anderson', who is described as 'raving around' (as cited in Estes, 1988, p. 182 n. 33).

This tale does have striking precedent, however, although not from New Orleans. Consider this account, dated 1858, and recounted in one of the early Spiritualist periodicals from the North:

> Mr. Conklin was at Washington. One of his children [in New York City] was taken ill, and attended by Dr. Gray, who considered the case so critical, that he advised the family to telegraph for Mr. Conklin. The party charged with this mission, on his way to the magnetic telegraph office, stepped instead into the office of the *Christian Spiritualist*... where a public circle was then in session ... [The medium] asked if there was any spirit then present who would take the message in advance of the telegraph. 'Black Hawk', the Indian chief, announced himself, and volunteered to do so ... Meanwhile Mr. Conklin was in his room at Washington ... reposing upon a sofa, when he suddenly felt impelled to get up and go to his table. There the influence came upon him, and he immediately wrote through his hand: *Go right home. Your little boy Sammy, is very sick. Black Hawk.* (as cited in Hardinge, 1869/1970, p. 267; italics mine)

The common motifs of the sick child, otherworldly communication, and in this case, the notable appearance of Black Hawk suggests that this narrative is a likely antecedent for the legend of Leafy Anderson's court case. It is vital to note, however, the significant transformation this narrative undergoes as an origin myth for the Spiritual churches in New Orleans. In its latter versions, the tale becomes more than a demonstration of spiritual presence and power through the revelation of a child's illness. Instead, it is a story about the *right* to parade, to process, to hold public worship and barefoot candle drills in the street. It is a story about inhabiting, possessing, and reclaiming, if only for a time, those frontier spaces of the city.

Spiritual church folks in the present day use urban geographic metaphors to describe the numinous power of the Indian spirits, in particular, their ability to 'make a right of way' for the devout through 'rough' moral territory. In much the same way that ornamental weapons are brandished by the Spy Boy in an Indian Gang, to clear the streets for the Big Chief, toy hatchets and tomahawks are brandished by Spiritual church folks during communal or personal ceremonies for the Indian spirits: to 'cut a right of way' for the individual through the obstacles that beset the everyday life. Framed by these narratives and ritual practices, the ability to move unhindered through the streets becomes, not a matter of parade licenses and social standing, but of moral uprightness and spiritual connection.

In addition to well-known communal legends like the one described previously, individual congregants and clergy-folk in the Spiritual church have personal narratives and histories that further establish the Indian spirits as the unseen forces controlling the streets. I asked Bishop Daniel Jackson, pastor of St. Daniel's Spiritual Church on Amelia Street to describe his personal relationship with Black Hawk, he began by asserting the Spirit's power over the legal system, then immediately shifted his characterization to one involving the ability to traverse the city streets.

'I use him for court cases', Bishop Jackson told me affably, 'I use him as my special protector: a shield and a guide'. Elaborating on themes of protection, the Bishop continued:

> I use him for my family ... I have children. You know, like whenever we go up and down dangerous highways. See I walk late at night ... I don't have a car and if I stay here 'til about 11:00, well then I'll have to walk quite a ways to the streetcar and then to my house. [Black Hawk] will work with you as a guide. Like you might be walking and spirit will say, 'Go left here', and you might be thinking, 'why should I go out of my way?' You were taught that the shortest distance between two points is a line ... but somewhere up a-head, spirit can see danger and you might not see it. That's what I mean by 'a guide'. (D. Jackson, personal communication, August 1997)

Bishop Jackson continued his description of Black Hawk with an account of the Spirit's very real presence on those 'dangerous highways':

> See spirit can take any form, to work with you. I remember one night I was just going home ... and out of no-where there was this big dog, [it] appeared and I started to run. And then when I thought about it, I said, that was the spirit of Black Hawk, see ... my guide. So I continued walking along one side of the bank, and the dog was walking along the other side. And out of nowhere, out of the blue, there came this man walking in front of me. And the dog went charging straight for the man, and he though it was my dog, even though I had no idea where the dog came from. And when I got home the dog was nowhere to be found. And I thought: 'That was Black Hawk ... that was my guide'. (D. Jackson, personal communication, August 1997)

Reverend Jules Anderson, a church minister with a particular affinity for Black Hawk likewise experienced an apparition of Black Hawk on the streets of the French Quarter. As he described this encounter to Jason Berry (1995):

> I was walking down the street in the French Quarter and this big Indian guy just came to me, in the spirit. It wasn't a real man. He was an Indian but he was dressed in ordinary American clothes, and he walked to me and folded his hands, like we see the statue of Black Hawk in Illinois. And he just walked right into me, and it was like a possession type thing I went into. And it was from that day on that I began to have this personal relationship with the spirit of Black Hawk, and I started to go around to different churches and other leaders told me I had the spirit strong. They said Black Hawk would work with me beautifully. (p. 19)

Whether manifesting as the disembodied voice of a kind of spiritual GPS system, in the form of a spectral dog, or as an apparition 'dressed in ordinary American clothes', the Indian spirit is accorded control of the streets: 'dangerous highways' and boulevards of the *Vieux Carre* alike.

The streets, as urban 'frontier space', are the natural environment for manifestations of the Indian spirits, whether metaphysical, as in the examples above, or mimetic, as in the performances of the Indian tribes. It must be understood, however, that these singular moments of expression in contested territory are fueled by continual, everyday interaction in domestic, or at least, *interior*, spaces. Long before a tribe member dons his suit and goes outside on Mardi Gras morning, he spends long hours each day 'entertaining' the Indian spirits, sewing and beading intricate expressions of Indian imagery and iconography. He may spend up to 12 hours on a Sunday at various 'Indian practices' with members of his gang and others, learning the songs, the dance steps, the secret language of challenge and confrontation. Likewise, while Spiritual church folk speak of encountering the spirits in dramatic visionary experiences on the streets of the city, much of their daily ritual work with the Indian spirits involves prayer, meditation, and other forms of spiritual *work* performed *inside* the church itself, or at private altars in personal domestic space. Within the sanctuary of the church itself, Spiritual folk use rituals of procession to challenge the rigid demarcation of inside/outside and re-imagine histories of space as well as of time.

Bringing the outside in: Holy Thursday procession at Israelite Divine Spiritual Church

Located at 3000 Frenchman Street, Israelite Divine Spiritual Church boasts a considerably larger church building than most other Spiritual churches, and its correspondingly larger worship space is graced with a wide array of striking altars dedicated to important saints and spirit guides. To step through the sanctuary doors at Israelite is to find oneself completely surrounded by images of the divine. The Black Hawk altar in the right rear corner of the church (on one side of the sanctuary door) is a large and well appointed one, boasting several Indian statues (including a beautifully painted bust of Black Hawk), a large St. Michael, spears, a peace-pipe, and a portrait of Leafy Anderson, along with ubiquitous candles, crucifixes, flowers, and vessels of clear water. Another Indian altar, this one dedicated to the mysterious spirit called 'Uncle Bucket' is located near the front of the church, to the right side of the main altar, just outside the door to the sacristy. The spirit to whom this altar is dedicated is variously described as a warrior or soldier and is also invoked for financial blessings (Hurston, 1931; Wehmeyer, 2000). Uncle stands guard over the inner-most sanctuary of the church (the room where bishops don their robes, where incense, oils, and other sacramentals are kept, and where the money offerings made by the congregation are locked up after a service.) This Indian altar reminds us that there are doors behind doors, mysteries within mysteries. To the left of the main altar is St. Joseph's altar, situated atop a carpeted staircase, with a large statue of the saint bearing his carpenters tools, aglow with the ruddy light of red pillar candles placed about his feet (as well as from red tinted electric lights that shine on this particular altar). Across the aisle (camera right) is a matching altar for St. Raymond lit in green, and bearing a statue of the saint holding a martyr's palm. Numerous green jar candles, lit and unlit, cluster about this saint's feet. I've been told St. Raymond is a 'good saint for financial blessings', and I think it significant that his altar hovers over the Uncle Bucket shrine, an angelic, saintly force watching over the more elemental, and more overtly operative spirit of Uncle below.

To the left of the aisle by the main doors, opposite the Black Hawk altar, stands a composite altar featuring St. Patrick, St. Anthony, and a small, generic Virgin Mary. St. Patrick's statue is particularly imposing: life size, dripping with Mardi Gras beads that hang around his neck, outstretched hands, and the crook of his bishop's crosier. At his feet sits a foil wrapped brick (brick dust is a powerful purifying and spiritual prophylactic agent in the pharmacopeia of Conjure work). During most of the public services I attended at Israelite, worship was largely focused on the main altar (the 'Throne of Grace' in Spiritual church parlance) and on the pulpit where the bishops and deacons preached. In 1998, on Holy Thursday, the feast commemorating Christ's triumphal procession into Jerusalem, I attended a service during which all of the altars at Israelite were engaged and *worked* through a simple but powerful act of ritual procession:

It is a damp Holy Thursday evening on Frenchman Street. My colleagues and I, here in New Orleans to attend the ACASA triennial conference, have arrived only a few minutes early for the weeknight prayer service at Israelite Divine. We pick our way quietly through the darkened sanctuary, towards the pews at the front of the church. One of the deacons kneels at the 'Throne of Grace', and prays privately, seeking inspiration for the coming service. Her kneeling form is lit only by a few flickering candle flames. As the official time for the service approaches, then passes, only a few more congregants have entered in silence and taken their seats. I answer a few whispered questions from the folks I've brought with me. No, it's not unusual for things to start a little late. Yes, I'm surprised at how few of us there are here. The electric lighting in the church brightens, and moments later, Bishop Oliver Coleman strides up one of the side aisles into the sacristy. He emerges a short time later in impressive robes, bearing a smoking censer. The heady scent of a fruit-based incense (pineapple? tangerine?) rapidly fills the sanctuary, there is a further moment or two of silence and then the service begins. The evening initially follows a familiar pattern. We are led in successive rounds of prayer, testimony, and meditation on scriptural passages directed by the deacons as the spirit moves them. The service seems a bit subdued, however. Our responses of 'Amen' and 'Yes, Lord!' sound perfunctory, almost half-hearted. There are precious few of us assembled for the service tonight. When the deacon instructs us to raise our hands to 'receive our bishop' who will 'come to the altar in his own way', less than a dozen weary arms go up. Oliver Coleman, whose rank as bishop marks him as a skillful orchestrator of spirit, has no intention of allowing the service to continue in this vein. He begins his portion of the proceedings by gathering the 9 or 10 of us present into a circle in the central aisle.

'C'mon folks, let's join hands in a circle … let's come together to pray at the altar. Everybody up!' he says jocularly.

Holding hands, we bow our heads as Bishop Coleman refills the censer and places it in the center of our prayer circle. The energy in the room shifts dramatically, becoming tighter, more intimate. I am sharply aware of the physical sensations of this place, this ritual. The incense tickles my nose … I feel the pressure of my neighbor's grip on my hand and I wonder if she notices my hand is sweating. We are all definitely more present than we had been before. As the sweet, thick smoke from the censer spirals upward, we pray and sing. Then Coleman speaks: 'Al-ll-ll right!' he exclaims, 'we're going to march! Three times, now. Then back to the altar'. Someone takes up a tambourine and the rest of us start clapping. The Bishop begins singing 'When the saints go marching in'. He leads us counter-clockwise around the perimeter of the church passing by each of the many altars gracing the sanctuary. St. Joseph looks benevolently down from his niche above us. Here is St. Patrick, crosier draped with green and white Mardi Gras beads, beside him a smaller St. Anthony, together with blessed Virgin. On the other side of the aisle we reach the Black Hawk/St. Michael altar, with its profusion of spears and feathers, and its glittering rack of red candles. As church members pass by these altars, they knock resoundingly three times on each one, honoring the saints and rousing them to action. We circle under St. Raymond, past the main altar again, and continue our march. On the next rotation, I make a special trip to knock on the altar of 'Uncle Bucket' (hidden as always in shadow near the sacristy door). Uncle's shrine is lit by single blue jar candle that flickers fitfully as I pass. On our final round, the singing has grown noticeably more passionate and robust. Bishop leads us back into our circle. One by one we 'smoke' the soles of our feet in the column of swirling incense, then brush the smoke over our heads and hands, and along our arms. We clasp hands again, eyes and faces shining through the smoke.

This ritual procession past the altars provides a key to understanding several different layers of Spiritual church belief and practice. On one level, we have a ritual comparable to the Catholic stations of the cross. The procession through the church brings us into successive contact with the saints and spirit guides, with each of God's individual messengers. Participants have the option of 'working' (through knocks and silent prayers) whichever of the saints they choose, as the entire array of sacred powers is presented to them during the procession. It is a pilgrimage in miniature, with congregants able to avail themselves of the entire church pantheon in the space of a few minutes.

While the operant, pragmatic aspects of a given saint (the specific blessings he/she confers, what he/she is *used* for) are usually foregrounded in the way a church member describes that saint ('I use him for court cases'; 'We call on him for financial blessings') it is essential to remember that many of the saints are also *ethnically* marked. St. Joseph, for example, is characterized as an 'Italian saint'. The specific rites associated with him (like the distribution of 'lucky beans' at his feast) are developments of comparable practices observed in New Orleans' Italian community (Estes, 1987). Likewise, St. Patrick is associated with the city's Irish community. Black Hawk, though ostensibly Indian, is the divine representative of New Orleans African-American community. That he is claimed as a definitively 'black' saint is borne out by the testimonies of a number of church members, past and present. Recall Rev. Jules Anderson's statement as quoted by Jason Berry (1995):

> I feel that Black Hawk has always been a part of the black spirit', explained Anderson. 'We was took from Africa and that left us – we didn't hear any of those drums. It was the Indians that brought us back to the drums and the music. (p. 20)

The ethnic identities of the various saints in the church also evoke associations with local geography. In our march past the altars of Black Hawk, St. Joseph, and St. Patrick, we are traversing the city's various ethnic neighbourhoods as well. The arrangement of altars at Israelite Divine thus establishes a microcosmic map of the ethno-spiritual territory of New Orleans. As we have already seen, the territory itself – the street *outside* the church – occupies an ambiguous position in the lives of the church members. These are the streets and neighbourhoods of home, avenues of commerce and opportunity. Yet as Bishop Jackson suggested, they are frequently dangerous and inhospitable; particularly to women and men trying to live what the churches would consider 'a spiritual life'. During Sunday 'practices', the Indians too recreate the exterior urban topography in interior space. Within the confines of local bars and social clubs, Indians line up to create an 'alley' along which Spy Boy, Flag Boy, Wildman and Big Chief will dance to meet visitors in mock confrontation . . . with their bodies the Indians metaphorically and viscerally bring the streets *inside*, in much the same way that church folks do with their altars.

Processions through these streets, processions that both reinforce and challenge established ethnic or class boundaries, are absolutely central to the cultural and civic identity of New Orleans. With its Carnival Krewes, 'Second-Line' societies, and jazz funerals, New Orleans is indeed a city constantly in procession. Our march around the altars of Israelite resonates with all of these processions, but perhaps most strongly with that of the Mardi Gras Indians.

When the congregation of Israelite Divine marches past the altar of the Irish St. Patrick, knocking to secure spiritual favor, they are also ritually reclaiming the Irish Channel, those two blocks of Adele Street between Tchoupitoulas and St. Thomas, that were once the site of brutal ethnic warfare. Armed with tambourine and holy song this procession marches past *every* altar, knocks on *every* door; asserting with each knock that no spiritual resource is denied them, no territory is beyond their reach.

On the corner of St. Claude and Governor Nicholls, New Orleans, Mardi Gras Sunday morning, 2006

'Oh yeah . . . we're still havin' church!' Bishop Coleman tells me, a trace of defiance crossing his otherwise bemused features. I'm standing with the Bishop outside the Backstreet Cultural Museum on St. Claude Avenue, in the Tremé district of NOLA. It's my first trip back since Hurricane Katrina, a fact of which I'm somewhat ashamed and embarrassed. I'm here to see what Mardi Gras will look like this year, and specifically, what role the Indians will play in the festivities. This has been a harsh and challenging year for the Indians as a whole. In March 2005, the Indian tribes suffered a violent disruption by New Orleans Police of their traditionally festive and peaceful gathering on St. Joseph's Night. Although prior relations between the Indians and the 6th District branch of the NOPD had been cordial, on this night in particular, the customary 'rite of territory repossessed' met with startling resistance. According to an article in the *New Orleans Gambit Weekly*:

> Consistent reports came from Indians and from spectators who'd gathered Uptown near A.L. Davis Park to watch the century-old St. Joseph's night Indian celebration. NOPD officers, they say, drove at high speeds on crowded streets, used foul language, and treated people roughly. One big chief's son, who alleges that he was manhandled by police officers that night, suffered tissue damage in his arm and is facing surgery.
>
> Among the Indians in the streets were highly respected chiefs who have celebrated St. Joseph's night in their handmade suits for more than 30 years. Old or young, Indians say they were given two choices: 'Take off your f – ing feathers or go to jail'.
>
> * * *
>
> Each group of tall feathers was followed closely by NOPD squad cars, which herded the Indians along . . . sirens wailing, buzzers pulsing, loudspeakers blaring – 'Get out of the street. Clear the street'. [This situation was observed by John Ellis, whose granddaughter Bryian marches as Spy Girl for the Wild Tchoupitoulas.] (Reckdahl, 2005, 29 March)

In June 2005, at a city council meeting convened in part to address the racial and socio-cultural implications of that St. Joseph's Night confrontation, the Indians' most respected Elder, Big Chief Allison 'Tootie' Montana of the Yellow Pocahontas, died of a heart attack while speaking from the podium in defence of the Indians and their traditions:

> Anyone who knew Montana knew him as a man of history. The big chief of the Yellow Pocahontas used the past to put the present into its proper context. Indeed, this was not the first time that the Indians had been mistreated by police, said Montana, 82, as he stepped to the podium and called the other chiefs in the room to stand behind him. Resting his left hand on the railing in front of him and gripping the microphone with

his right, Montana told of a time years ago when he and the Yellow Pocahontas were treated roughly by the New Orleans Police Department (NOPD). 'I want this to stop', he said, looking directly at the council and top NOPD brass.

Then he fell.

His son, Big Chief Darryl Montana, held his father while officers rushed forward and began CPR. About 10 minutes later, paramedics rushed in. Councilman Oliver Thomas adjourned the meeting and led the room in prayer, asking God that this moment bring everyone together. Then chiefs in the room began singing the sacred Indian chant, 'Indian Red'. As paramedics carried the chief of chiefs up the aisle, the room was silent except for the hushed singing of the chiefs and the voice of his wife, Joyce, who followed the stretcher, telling her husband she wasn't ready for him to leave her.

Montana was pronounced dead of a heart attack a few minutes later at Charity Hospital. 'Tonight, a legend left us', said one man, as he stood outside Charity's emergency-room door. When the news cameras came, people looked into the lenses and talked about Tootie Montana, the legend, who had died the way an Indian should – fighting for what's right and surrounded by his chiefs. (Reckdahl 2005, July 5)

And two months later came Katrina. And lightning and thunder, indeed, put the world onto wonder.

With the rest of the country I watched in dismay and despair as the levies ruptured and the water rose. I tried desperately and worriedly to get through to the

Figure 1. Mardi Gras Indian and Northside Skull and Bones masker embrace outside the Backstreet Cultural Museum, New Orleans. Photo by Stephen C. Wehmeyer, 2006.

cell-phones of my friends and associates from the Spiritual church communities, finally reaching Bishop Coleman weeks after the crisis, when he'd successfully removed his family and a few belongings to safer ground in Baton Rouge. When the waters receded, like everyone else I saw the photographs of the devastated 8th and 9th wards, and I was deeply moved by those houses which bore, alongside chilling FEMA tags, the bedraggled remains of Mardi Gras Indian suits, nailed to front doors and balustrades in mute, but indomitable, witness.

Now I'm back in New Orleans, on a balmy, early morning, waiting outside the museum to see if the Indians will indeed come to take back the streets. Many of the Spiritual church folks are already back. Some, like Coleman, are living in FEMA trailers, as they struggle to clean and repair flood-damaged homes. Many still lack power and clean water. But at Israelite Divine, at least, they're still having church. Bishop Coleman introduces me to a friend from the Northside Bone Gang, a cheery fellow garbed in a chilling costume. His face painted like a skull, he wears black-sweat-suit adorned with bones and a hand-illustrated apron bearing the ominous legend 'You Next!' He's living outside New Orleans at present, but wants to come home. 'Are you going to come back?' asks a visiting reporter. 'We wanna come back!' the Bonesman answers. 'Give us water, give us power . . . sure we'll come back!'

Figure 2. 'Red Hawk Hunters' Spy Boy at Claiborne Street overpass. Photo by Stephen C. Wehmeyer, 2006.

While we chat, an Indian, garbed in black and gold has arrived at the edge of the crowd. The Bonesman sees him and crows with delight, then issues a ritual challenge: 'I'm a Bone man, from the Bone Gang! Northside! What's your name? What's your Gang?'

We all make way and the two face off, as the Indian responds with a softly-yodeled 'Heeeeyyyyy-ey!' Bobbing and weaving up to one another in mock combat, the two embrace warmly, the Bonesman welcoming the newcomer. It's impossible not to read this reunion metaphorically: Death has held New Orleans' streets for these long months, awaiting the Indians' return (see Figure 1).

Coleman and I mill around for a while after this encounter, then the Bishop excuses himself, and I wander over to the Claiborne St. overpass, an important meeting ground for the Indian tribes in recent years. Once a prosperous African-American thoroughfare just northwest of Louis Armstrong Park, the construction of the 10 freeway overpass ruined local businesses and irrevocably changed the neighbourhood. Many of the Indian tribes end their Mardi Gras processions here: beaded boots reclaiming this hotly contested turf, which could certainly be considered one of Bishop Jackson's 'dangerous highways'. Today, however, there's a block party in full swing (sponsored by local KMEZ radio station) and despite the afternoon heat, the place is jumping harder and rowdier than anything I've seen this

Figure 3. 'Won't kneel! Won't bow down!' Indian with flood damaged shield under Claiborne Street overpass. Photo by Stephen C. Wehmeyer, 2006.

weekend in the 'Quarter. I'm buying an illicit beer from a wary looking guy with a full cooler when the shouting of the crowd around me threatens to drown out the band. The Creole Osceolas are marching past: my first tribe of the afternoon. Most are unmasked, their fine costumes lost in the relocation, but a few carry feathered flags and standards, and all wear commemorative t-shirts or sweatshirts. Costumes or no, they're marching anyway. The Yellow Pocahontas follow soon after marching to honour Big Chief Tootie, his picture carried high in the center of a blue-plumed standard.

Not long after, I turn to see the Spy Boy for the Red Hawk Hunters sallying past the pillars of the underpass, clearing the way for the tribe to follow (Figure 2). The flag boy follows, in full yellow costume and full character, fiercely challenging anyone who steps in his way. Red Hawk Hunters' Big Chief, in a massive cloud of crimson feathers, isn't far behind. Just beside him dances a fellow whose only costume is a single pink plume in his hat. In one hand, however, he brandishes the scraggly remains of a feathered

Figure 4. In Memoriam: Donny Harrison Snr. – Big Chief of the Guardians fo the Flame. Photo by Stephen C. Wehmeyer, 2006.

shield, clearly salvaged from the flood but wielded as proudly and defiantly as any of the unharmed finery adorning his Indian brothers (Figure 3). He glares at me and my camera, and thrusts the shield skyward, then dances away, unconquered.

On my way back to the car, I pass a house where a poster-sized photo stapled to a wooden staff rests propped against the front stoop. Detritus from this morning's procession? Or something else? Carried that morning in procession by the members of the Congo Nation, a tribe whose Big Chief, jazz musician Donald Harrison, Jr., is devoted to expressing the African roots of Black Indian tradition, this standard now serves as a silent but active shrine for the spirit of Harrison's father, Donny Harrison, Sr., Big Chief of the *Guardians of the Flame.* Beneath the sign, a lump of self lighting charcoal smoulders atop a half-brick. A few remaining lumps of resinous incense send wisps of aromatic smoke curling around the wooden staff. 'Entertained' and sanctified with the same paraphernalia used on Indian altars in the Spiritual church, in the final hours of Mardi Gras, this honoured Chief keeps watch over the quiet street (Figure 4).

Clearly, the effects of Katrina and its aftermath are still shaping New Orleans in innumerable ways, many as yet undetermined or little understood. Devastation, relocation, administrative breakdown and corruption, a 'new diaspora' of the newly displaced: we are only beginning to see what such elements will do to African-American cultural institutions like the Spiritual churches, and the Indian tribes. For now at least, in prayer or in procession, Indian icons and imagery – the embodiments of Indian spirits, still stand for many New Orleanians as powerful signs of something in the soul that refuses to kneel and will not bow down (Figure 5).

Figure 5. Future site of the Tootie Montana Institute at Dumaine and North Claiborne, New Orleans. Photo by Stephen C. Wehmeyer, 2006.

Acknowledgements

In the ritual practices of the Spiritual church, before testifying, it is customary to give honour to 'God the Father, his Darling Son Jesus', thence to the Archbishop, and any other high ranking clergy in attendance. In this tradition, I must give honour to the men and women of the Spiritual church (notably E.J. Johnson, Edmonia Caldwell, Oliver Coleman, and Daniel Jackson) who warmly and openly shared their beliefs, ideas, thoughts and traditions with me. I give honour as well to those fallen warriors, Big Chief 'Tootie' Montana and Big Chief Donny Harrison, Sr. (whose rendition of 'Indian Red' I heard at Vaughn's tavern in the Bywater on my very first night in New Orleans).

Note

1. Portions of this essay have previously appeared as part of my doctoral dissertation *Red Mysteries: 'Indian Spirits' and the Sacred Landscapes of American Spiritualism* (Wehmeyer, 2002) and in the article 'Indians at the Door: Power and Placement on New Orleans Spiritual Church Altars' (Wehmeyer, 2007).

References

Berry, J. (1986). *Up from the cradle of jazz*. Athens: University of Georgia Press.
Berry, J. (1995). *The spirit of Black Hawk*. Jackson: University Press of Mississippi.
Blank, L. (1978). *Always for pleasure* (Motion Picture) USA: Flower Films.
De Caro, F. (1978). Review of *The Black Indians of New Orleans*, by Maurice A. Martinez and James E. Hinton. *Journal of American Folklore, 91*, 625–627.
Dollis, B. (2002). Mighty mighty chief [Recorded by Bo Dollis and the Wild Magnolias]. On *30 Years... and still wild!* [CD]. New Orleans: Louisiana Music Factory.
Estes, D. (1987). Across ethnic boundaries: St. Joseph's Day in a New Orleans Afro-American Spiritual church. *Mississippi Folklore Register, 21*, 9–22.
Estes, D. (1998). 'Hoodoo? God Do' African American women and contested spirituality in the Spiritual churches of New Orleans. In E. Reis (Ed.), *Spellbound: Women and witchcraft in America* (pp. 157–182). Wilmington: Scholarly Resources.
Hardinge, E. (1970). *Modern American Spiritualism*. New York: University Books. (Original work published 1869)
Hurston, Z.N. (1931). Hoodoo in America. *Journal of American Folklore, 44*, 317–417.
Jacobs, C.F., & Kaslow, A. (1991). *The Spiritual churches of New Orleans*. Knoxville: University of Tennessee Press.
Lipsitz, G. (1988). Mardi Gras Indians: Carnival and counter-narrative in black New Orleans. *Cultural Critique, 10*, 99–121.
Martinez, A., & Hinton, James E. (1977). *The Black Indians of New Orleans* (Motion Picture) USA: Chalmalma Media Institute.
Nunley, J., & Bettelheim, J. (1988). *Caribbean festival arts: Each and every bit of difference*. Seattle: University of Washington Press.
Orsi, R. (1999). Introduction: Crossing the city line. In R. Orsi (Ed.), *Gods of the city* (pp. 1–78). Bloomington: Indiana University Press.
Reckdahl, K. (2005, 29 March). St. Joseph's Night gone blue. *New Orleans Gambit Weekly*. Retrieved October 13, 2007, from http://bestofneworleans.com/gyrobase/Content?oid=oid%3A34185
Reckdahl, K. (2005, 5 July). Chief of chiefs. New Orleans Gambit Weekly. Retrieved October 13, 2007, from: http://bestofneworleans.com/gyrobase/Content?oid=oid%3A34791
Roach, J. (1992). Mardi Gras Indians and others: Genealogies of American performance. *Theatre Journal, 44*(4), 461–483.
Smith, M.P. (1992). *Spirit world*. Gretna, LA: Pelican Publishing.
Smith, M.P. (1994). *Mardi Gras Indians*. Gretna, LA: Pelican Publishing.
Smith, T. (1994). *Conjuring culture. Biblical formations of Black America*. New York/Oxford: Oxford University Press.

Thompson, R.F. (1993). *Faces of the gods: Art and altars of Africa and the African Americas.* New York: The Museum for African Art.
Wehmeyer, S. (2000). 'Indian altars of the Spiritual church – Kongo echoes in New Orleans'. *African Arts, XXXIII*(4), 62–70.
Wehmeyer, S. (2001). 'Response to [Judith] Bettelheim'. *African Arts, XXXIV*(3), 91–92.
Wehmeyer, S. (2002). *Red mysteries: 'Indian' spirits and the sacred landscapes of American spiritualism.* Doctoral dissertation: University of California, Los Angeles.
Wehmeyer, S. (2007). Indians at the door: Power and placement on New Orleans Spiritual church altars. *Western Folklore, 66*(1), 15–44.

Bahian carnival and social carnivalesque in trans-Atlantic context

Piers Armstrong

Department of Modern Languages and Literatures, California State University, Los Angeles
International Center, UCLA

This article outlines strands of carnival theory from Bakhtin to Black Atlantic and East African post-colonial scholarship, then describes the carnival of Bahia, Brazil. In New World carnivals, race is a critical rhetorical category in the exchange between everyday authority and the subaltern cultures whose energies manifest in the event. The most racially self-conscious of such carnivals, the Bahian carnival is generally considered a beacon of Afro-centricity and constant aesthetic evolution, and associated with resistance to patriarchal order. Here, as elsewhere, however, carnival praxes can be read as a reflection of social conflicts and collusions, but not as a transparent translation of material relations. As well as individual affirmation and traditional group identities, carnival now fosters consumer cultures. As to its symbolic logic, progressive rhetoric is contradicted by the surprising viability of a conservative agenda in which 'dissidents' often collude. Bahian carnival is less an inversion of order than an intensification of an everyday culture which has already assimilated carnival motifs and values. The temporal and spatial boundaries between these zones are unusually open. The composite middle-ground is here termed the 'Bahian carnivalesque'; though less conspicuous, this social sensibility of tolerance, with all its ambiguities, is more extraordinary than the carnival itself and may better survive commodification.

Introduction

Tensions between socio-racial elements in colonial and post-colonial societies have historically been given indirect expression in popular culture, particularly carnival. In viewing carnival performative genres as a social logic, we must consider both the general mechanics of mimesis – the interrelation of art and social life – and also contextual questions: how the social logic of these carnivals differs from the original European carnivals; how the colonial and post-colonial phases of carnival history differ; and to what extent all contemporary carnivals are affected by globalization and other universalizing trends.

One relatively consistent aspect of carnival theory is the centrality of the dialectic relation between a dominant cultural order and the subaltern masses. In Bakhtin's study of medieval Europe, the dialectic was between the top-down moral and political order and the popular classes. The masses used carnival to symbolically vent and appease socio-political frustrations and to express spiritual notions not condoned by orthodox Christianity. In his account, the coupling of motifs of order and subversion, conceived as inverses or opposites, generates a parallelism which in fact unites them in an organic whole: the society is able to symbolically purge its demons and renew and restore itself, ultimately preserving the synergy between contrastive energies (Hirschkop, 1999). This view is reiterated by more recent theorists such as Allesandro Falassi, who writes about festivals generally: 'the primary... function of the festival is to... renew periodically the life stream of a community by creating new energy, and to give sanctions to its institutions' (1987, p. 3).

Carnival can also work the opposite way and be harnessed to the interests of civic adhesion, particularly in the context of Latin American nationalist populism. Popular expression in these carnivals has been analyzed both positively, as a rare wedge into the public sphere affecting the socio-political agenda in a 'bottom-up' way, and negatively, as a superficial therapy for the dispossessed.

The agency of the state may go even further. Contemporary critics, notably prominent Africanists drawing on carnival theory in the abstract and speaking about societies which do not celebrate carnival *per se*, suggest that the oppressive regimes of many post-colonial states constitute a carnival logic in ethical and spectacular terms. Through ostensive displays and public charades with which peers must play along, dictatorial heads of state impose a civic narrative which is collectively corroborated. Such coercion may have crude fascistic overtones, but in this post-colonial critique the Bakhtinian carnivalesque is suggested in a more particular way because of the centrality of grotesque bodily aspects and sheer excess both in official motifs and in the responsive humor which consequently circulates in the citizenry; what begins as intimidation and absurdity devolves into sense and sensibility. In this scenario, a 'top-down' agenda 'carnivalizes' society. On the other hand, a rhetoric of carnival is adopted by some metropolitan scholars as a theoretical buttress to 'bottom-up' agendas of healthy subversion (for example, see the discussion of 'tactical carnival' by L.R. Bogad in the present volume).

Carnival is thus both a real event and a putative social aesthetic distilled by theory and applicable to societies with or without carnival, and figures both as a conduit for subversion and as a lubricant for hierarchical control. The characterization of carnival as a special time hinges on the related characterization of everyday order. In the contemporary perspective, as distinct from Bakhtin's, where carnival is linked to subversion it is cast as challenging a potentially unstable everyday order, whereas as a hierarchical reinforcement it merely reflects a more implacable order. Carnival license and everyday order thus exist in a symbiosis which is very ambiguous in terms of political agency. If we assess this symbiosis not in terms of contrast but rather of complementation, it remains difficult to map how and when the boundaries are traversed, and to distinguish cause from effect (carnival precipitating, or rather reflecting tension and potential change).

In the literature, Bakhtin's pioneering analysis of medieval carnival logic remains the most consistently referenced theoretical frame. As case studies, however, the living carnivals of the Americas have gained prominence in diaspora studies as highpoints of the popular cultural calendars of their respective societies. In the latter context, carnival has generally been viewed as an existentially distinct moment. This essay is ultimately concerned with the overlap of carnival and everyday order. It draws on classic carnival theory in contrasting the carnival and everyday orders as poles, and then probes the derivative terms of 'carnivalization' and the 'carnivalesque', as intermediary categories. Theoretical meanings for these two terms vary considerably. They are used here with some particularity: 'carnivalization' refers to material events which lie outside carnival but are organized like it; 'carnivalesque' refers to abstract values pertaining in everyday life which suggest a cultural logic or sensibility akin to the 'carnival aesthetic'.

Bahia (and more particularly Salvador, the former colonial capital of Brazil) is a particularly dynamic arena in which to study the polyvalent play of carnival. Afro-Bahians have traditionally used the license of carnival to affirm ethno-social identity with an Afro-centric dissidence distinct from other Brazilian carnivals. However, the negotiated spaces of the carnival also reiterate the privileges of the Euro-Brazilian oligarchy. Both Afro-centric articulations and social conservatism are features of Bahian carnival, compounding the political ambiguity of the inherent ritual symbolism. Bahian carnival also presents great variation over time, and the volatility of its performative modes is a distinguishing feature relative to other carnivals. Bahian carnival tends to respond more promptly and conspicuously to change, to reflect evolving social agendas and consumer interests, and seems almost as susceptible to modernization as is everyday consumer culture. Complementing this receptivity to innovation, the relation between carnival and everyday culture is particularly intimate in the Bahian context. Temporal and spatial boundaries between carnival and everyday culture are relatively open so that the two domains are less polarized and their respective performative modes converge.

Two broad questions instigated this essay. The first is more political: what alternative social visions, interests and strategies pertain, and what is the balance between subaltern racial affirmation and commodification? The second – the relation between the symbolic logic of carnival and its subliminal presence in everyday civil society – concerns the social imagination. To what extent are carnival practices existentially unique? Are they a sublimation or a subversion of everyday embodied knowledge it? Does the conspicuous black inflection in carnival stand for socio-economic dispossession generally or rather a culturally particular, ethnic heritage? How does modernization affect these patterns?

The essay explores post-colonial currents in carnival theory to draw out certain aspects of the theoretical puzzle. Turning to real carnivals, the Bahian case is briefly compared to other prominent diaspora cases, and then investigated on its own terms, in historical perspective and in relation to ethno-social symbolism and political rhetoric. The final section considers how the findings in this case study resonate with the theories outlined.

Carnival and the carnivalesque

Each carnival can be seen as a repertoire of expressions whose deeper symbolic meanings derive from certain conditions in its anchoring culture. To the degree that these are analogous with the modes of the carnival itself, they can be called 'carnivalesque'. This is a slippery term, one often used in a loose way to speak of those life-situations or aesthetic arrangements for which carnival is a broad metaphor, frequently in cultural contexts independent of carnival. When invoked by carnival theorists, the term 'carnivalesque' may mean some of the following: the ludic (spontaneous and playful) behavior associated with carnival license when manifested in ordinary moments; philosophies of life which are influenced by ideas prominent in carnival, including masking or changeability; the idea of seizing the moment; sensuality and social communion; and, finally, the heterogeneous assemblage of ethnic and socio-cultural enclaves which interact performatively under the umbrella of carnival but also in a less formalized way in the everyday life of their respective societies.

The symbiotic relation between carnival and the general order does not mean that the aesthetic praxes of carnival directly reflect social reality or that the repertoire of imaginative roles in carnival reveals power relations. Carnival art tends toward encoding and toward being a separate, autonomous game which resists clear distinction between external reality and internal contrary desires. Like myth, the play of carnival may articulate, manifest or suggest issues in a colorful way, but without proving anything. On the other hand, as a prominent practice, and as a material reality rather than a jumble of symbols, carnival impacts society and is emulated in other domains of social performance, including politics. This sort of conspicuous carnivalization is a good base to start from, in assessing whether and how, at the more intangible level of cultural praxes and values, the consequent social character might be called *carnivalesque.* Both terms – carnivalization (material organization and praxes), and the carnivalesque (immaterial cultural consciousness) – imply a measuring of the overlap between carnival itself and everyday life, even though these are usually presented as contrasting and discrete domains.

Carnival feeds on various sorts of what the anthropologist Victor Turner (1969) called 'liminality', that is, the cathartic conjunction of contrastive categories: the consecrated and the profane, the ritualized and the spontaneous, domination and subversion. The central icon of carnival is the mask. As a hinge between an external and a hidden identity, the mask speaks to the importance to carnival of liminality. The external identity proposed by the mask points to an aspiration, to an ideal or otherwise desired identity, and alerts us to how identity is asserted in carnival principally through theatrical gestures, which are seen in a mirror of self-image as much as by onlookers. Liminality is lived out in carnival in a rhetorical cocktail which temporarily closes the gap between frustration and empowerment. Thus, despite the distinction between carnival and everyday life, the carnival aesthetic does not project us into a hermetic world of fantasy but rather into an aesthetically sharpened and socially rooted tension between the two domains (Turner, 1969, 1987).

Black carnivals of the Americas

In the diaspora societies of the Atlantic and the Americas, race is a critical category in the exchange between everyday authority and subaltern cultures. Ethno-cultural logics – the world-views, circumstances and praxes characteristic of certain communities – are often perceived in stereotypes, relevant despite cultural complexity and individual variation. At the crudest level, racial profiles link physical appearance to a determined quotient of social legitimacy and a supposed psychological make-up. Such perceptions have been expressed harshly in various legally-sanctioned inequities. While these were abolished as official distinctions through the liberation from colony status, a more persistent social logic underscores their enduring presence as cultural narratives pertaining in subjective perception, of self or of others, in affirmative or negative terms. Individuals are inevitably implicated in the web of such narrative logics in everyday social intercourse. Amongst these, racial profile is simply the most acute and reductive way for an individual's social interlocutors to categorize him or her. In short, race is an embodied mask whose code is determined by the viewer more than the embodier, and tainted by inequitable power relations (Aching, 2002).

Carnival creates a space for the disruption of narratives which reiterate everyday hierarchies, and the carnival aesthetic relates organically to counter-patterns indirectly articulating the counter-narratives of otherwise marginalized interests. But New World history also complements the logic of carnival: the process of creolization in the diaspora suggests the liminality of Europeans and Africans and their cultures, despite the hierarchic distance of the colonial order. The social narratives implicit in carnival play reinforce the motif of creolization, though perhaps more as a sublimated ritual experience than as a conscious discourse. Carnival is a stage for the affirmative representation of traditionally diverse socio-ethnic entities; it alludes to their conflicts, collaborations and collusions in everyday life.

While carnival underlines group identities, it is also a market for an emerging consumer culture. Assertions of traditional group identity are complicated by the difference of aspirations between individual consumers, and by circumstantial pressures in the form of the vested interests of performers and other suppliers of culture. The modernization of carnival enhances another traditional liminality, between collectivity and individuality. Under the umbrella of carnival, group versus individual agendas interact ambiguously, as do dissident groups' affirmations of cultural uniqueness and their very conventional behavior in securing material interests. Through the maze of carnival opportunities, the various interested parties negotiate both symbolic codes and material opportunities, as representatives and as agents, as the re-iterating players of carnival and as its re-drafters.

Postcolonial takes on the carnivalesque

The prominent carnivals of the Americas arose in independent states which had been colonial slave societies. Each presents particular features. Brazil and many Spanish American nations were staunchly Catholic colonies of the Iberian empires and won independence early in the nineteenth century. Trinidad, like various other Caribbean islands, experienced a rotation of metropolitan empires (in this case, Spain, France and Britain), gained independence well into the twentieth century, and developed its carnival during the transition. Haiti underwent a race-based revolutionary process consequent to the French Revolution and gained a more radical independence. Within the US, New Orleans is considered to be a uniquely Caribbean culture, which preserved nuances of colonial power relations in its complex codes of color and race-adscription (DaMatta, 1978/1991 provides a Brazilian perspective on Mardi Gras and Louisiana's relation to the rest of the United States).

The term 'post-colonial' supposes the survival of patterns of hierarchical power, albeit with a national cast. The heritage of hierarchy has generally meant political authoritarianism, as in many African, Asian and New World dictatorships, or a limited democracy, as in Mexico, Argentina and Brazil, for example (whether in the populist regimes of the mid-twentieth century or their current rewritings as more conventional western democracies).

All traditional carnivals were spawned in sharply hierarchic societies, and where they survive today are clearly less prominent in the more horizontal democracies of Western Europe and the US. Is there, then, an organic relation between carnival, hierarchy and illusory democracy in the Americas? Are these carnivals distractions or incentives to popular resistance, or both simultaneously?

Thinking past Bakhtin, Richard Schechner suggests that in the contemporary Americas 'the social "baseline" is democracy or the illusion of democracy' (2004, p. 3). Carnival 'plays out democratic illusions, giving temporary relief from the authority (if not oppression and downright tyranny) imposed in the name of "democracy"' (2004, p. 3). His view is more nuanced when applied to the concrete case of Trinidadian carnival, emphasizing the ambiguity of its performance, where 'a "forgiveness of illogic" allows contradictory values to be simultaneously expressed (2004, p. 6) – and the ambiguity of its political meaning, which, while managed by the government is also 'a dangerous, almost about to come apart, coalition of traditions and socio-political arrangements' (2004, p. 7).

Schechner's concern regarding Trinidad is to decipher carnival through performance theory. A related deep-reading approach to the existential dimension of carnival and other festivals is taken by the Rio-based anthropologist Maria Laura Cavalcanti, whose sources include Durkheim's writings on Australian Aborigines and the philosophical phenomenology of Merleau-Ponty. The main concern of a line of Rio researchers, including Muniz Sodré, Ana Maria Rodrigues and Maria Isaura de Queiroz, is the appropriation of black cultural capital by event managers (often Euro-Brazilian) in what is merely a variation on that society's everyday patterns of exploitation. In contrast, in works published by a think-tank named for the former president Getúlio Vargas, another line of Rio researchers stresses continuity over time and, implicitly, cross-class and inter-ethnic harmony (Rodrigues, 1984; Sodré, 1999; Queiroz, 1992; Cavalcanti, 1999).

A further feature of the modern democratic or pseudo-democratic context (and this time, the distinction is less important) is its reality as a popular market, an arena of economic opportunity, exchange, and negotiation between individual and group agents who are constrained by numerous circumstances but still autonomous. This is the spirit of much of the scholarly literature of New World carnivals. For example, another line of Brazilian scholarship, in Bahia, while preserving a sense of the hierarchical and race-coded nature of Brazilian society, looks at how the carnival industry affords opportunities for disenfranchised groups and individuals (Dantas, 1994; Fischer, 1996).

The lines mentioned exemplify successive views by the Left in relation to agency, with the earlier phase emphasizing exploitation of passive victims and the later phase seeking evidence of creative agency. While more attractive, the agency view – by now the dominant trend in the carnival literature – has its own problems. First, it is a structural position to which many North Atlantic scholars naturally tend because of their own need to avoid the dilemma of objectifying their objects of study. A second problem of the pro-agency position is that it is easily assimilated to neo-liberalist views of the vital role of small business entrepreneurship and to nationalist apologias whereby hierarchy is seen as a background rather than a foreground, as a legacy which is being unwritten rather than a proof of the persistence of race-based oligarchy.

These various lines of enquiry share a concern with the relation between carnival and everyday life. However, they do not engage the more difficult question of the carnivalization of everyday life or pursue notions of a carnivalesque character flowing to broader society from the carnival or from whatever deep need in the broader culture fosters the popularity of carnival. One of the most radical adaptations of Bakhtin's theory of the carnivalesque is that of the Africanist and post-colonial theorist, Achille Mbembe. In his 1992 article, 'Provisional notes on the postcolony', which focuses on governance in Cameroon, Mbembe draws on Bakhtin's notion of a subliminal discourse of class contestation inherent in the aesthetic categories of the grotesque and the bodily metaphors which pervade European medieval folk culture and are heightened in carnival. Mbembe argues that African postcolonial regimes, and thus the hegemony rather than the subalterns, adopts a carnivalesque discourse in its ostensive display, excess and focus on bodily metaphors in figures of speech. Mbembe characterizes an aesthetics of postcolonial power, and, recalling Foucault, explores the identity games in which individual citizens are obliged to engage, whereby they become complicit without really accepting the legitimacy of their rulers. As oversized, bullying and infantile personalities, Mbembe's African post-colonial rulers are reminiscent of the Ubu kings of the absurdist playwright Alfred Jarry (2003), suggesting a farcical and neo-fascist ambience which successfully infiltrates civic life and social intercourse.

The degree of real power exerted under such authoritarianism might well be compared to Latin American dictatorships but not to the contemporary democracies of Brazil or Trinidad, for example, regardless of their degree of 'illusory democracy'. As Mikael Karlstrom points out, Mbembe's view is top-down and pessimistic. Karlstrom notes Mbembe's rejection both of Bakhtin's notion of a Utopian aspiration in medieval folk resistance, and of the possibility in contemporary African postcolonies of a popular resistance to abusive leaders based on a substantial contrary political aspiration. Drawing on field studies in Uganda and providing detailed accounts of the performative mores of interactions between rural communities and state authorities, Karlstrom defends the existence of a 'popular political subjectivity [which] contains resources for the sort of critical consciousness of which Mbembe despairs' (2003, p. 61). In the face of a greater authoritarianism than is current in Latin American new democracies, Karlstrom still insists on the role of the people in articulating the terms of a progressive political project, no matter how frustrated by circumstance.

Applied to our scenario, Mbembe's view does not allow for the autonomous negotiation of official order by the groups and individuals who play carnival. What he does show is how a carnival aesthetic may permeate the everyday cultural order – admittedly, at a metaphoric rather than a literal level, since his extreme instance of carnivalization is consequent not to an actual carnival but to a perverse political regime. Working similarly with a metaphoric application of the carnivalesque, Karlstrom bolsters our faith in the pertinence of popular expression. Without proof, we might accept Karlstrom's optimism for no other reason than keeping good faith in popular sovereignty. Karlstrom's recourse to the rural domain and thus a more traditional framework is problematic, however, for the diaspora states of the Americas, in which carnival is a markedly urban phenomenon.

Where do the postcolonial theories of the carnivalesque, as well as Bakhtin, leave us overall, then, in relation to the contemporary carnivals of the diaspora states of the Americas? Do these cases validate Karlstrom's optimistic view of popular sovereignty? If we accept the notion that the subaltern side of a popular/elite duality gains an outlet in carnival, but in such a way that it may expend its own fire (and not lead to sustained conflict), then this seems a validation of the hegemonic order and the status quo. Indeed, the very occurrence of large-scale carnival is predicated on widespread acceptance of highly focused cultural conceits, and thus reiterates a bond which underwrites the established order. This social centering by carnival is described below as a 'centripetal' effect – drawing people together in an informal social contract – which may pertain even where dissident rhetoric offers contrary 'centrifugal' proposals intended to demystify official pretensions and generate a critical awareness among citizens. Secondly, while the motif of subversion in carnival pervades academic discussions, carnival also honors its presiding authority figures and allows for the reiteration of conservative messages. If the jestful polemics integral to the carnival ambiance provide license for the public ridiculing of figures of authority, the same process validates the regime's constant assertions that a substantial degree of democracy exists.

For many scholars, the agent eroding popular culture in carnival is not conspiracy by a political elite but commercialization – the deployment of carnival as a leisure commodity to be purchased by consumers rather than an intangible cultural game to be played by participants. But against this process of commodification in the transition from popular to mass culture, modern carnivals still need popular authentication. A completely commodified, rationalized or orchestrated carnival would lack flavor. Carnival must ignite the imagination of the disenfranchised masses to stimulate their enthusiastic participation, and in so doing it must speak to their cultural agenda.

This bottom-up constant is also complemented by a progressive top-down agenda which does not usually derive from the local political power. In their official procedures and tone, modern carnivals tend toward international (i.e. Western or United Nations) norms, which are very progressive relative to the local societal status quo. The political arguments presented in carnival often engage the 'politically correct' and are gradually assimilated into official discourse. In the context of the developing democracies of the Americas, carnival is an arena in which civil society expresses itself in a somewhat idealized or charitable form – with respect for the environment, valorization of subaltern communities, and so on. As a forum, it is receptive to the aspirations of the poor to full material and cultural citizenship. While carnival is co-opted as an apologia for the state, as a cultural praxis it draws on the popular, reflecting and thus acceding to popular demands – if not for immediate material reparations then for the normalization of progressive discourses or more widely representative images, many of which eventually become part of the political and cultural mainstreams.

Assessed empirically at the level of gross pragmatic effect, carnival presents contrary energies. On the one hand, it is a centripetal rite whose conservative effect in reiterating and implicitly celebrating existing order is greater than the contrary centrifugal effect of opening up space for dissidence. On the other hand, in the long term the conservative effect is strongly countered – not only by conspicuous and conscious dissidence, but also by the gradual assimilation to the national mainstream of imported references and local 'bottom-up' expressions. Whereas in medieval times carnival dissidence projected a temporary alternative which reverted to taboo in the everyday order, today carnival affirmations of popular will sketch out agendas which politicians must 'respect', or at least reiterate in campaigns. In short, the reading of carnival as an escape mechanism for the poor which is, in fact, strategically amenable to hegemonic interests does not allow for the possibility of the popular classes 'losing every battle and yet winning the war'. Just such a positive process does seem to pertain in the historical evolution of societies and the gradual shifts of relations between high and popular culture, warranting optimism that the people somehow do prevail.

This is nowhere more apparent than in the historical adventure of the popular cultures of the black diaspora, which presents case after case of repressed subaltern agendas gaining circulation and attaining legitimacy in the mainstream. A further feature of many of the black diaspora cultures is a legacy as prospective as it is retrospective, a capacity for self-reinvention which assimilates elements of mainstream modernity while preserving a distinctive ethno-cultural continuity.

Carnival is a notable venue for the display of this general continuity of black culture, but, more than this, as an arena, it presents affinities with the specific history of the Black Atlantic. Both carnival and diaspora culture derive from an ethnically-specific, popular ludic resistance to severely hierarchic societies of the past, yet readily use technology, digest the modern imagination, and reach beyond their base-communities to draw in a wide reception. Carnival's role may be to instigate or merely to reflect this process of successful bottom-up outreach.

The theory of carnival is as inconclusive as its object of study is ambiguous. As Karlstrom and Mbembe show, however, Bakhtin's theories of the carnivalesque can speak to post-colonial social hermeneutics even where carnival is not celebrated. Their studies are useful points of departure for comparative assessments across the Afro-diasporic world, whether of the carnivalesque in an abstract sense as it pertains in putative democracies, or of actual carnivals seen in social context.

Bahian carnival: basic format and comparisons with other New World carnivals

The basic format of Bahian carnival is live music in the street performed by popular local artists. They play on huge, slowly moving trucks called *trios elétricos* (or *trios*). The roof of the *trio* serves as a stage able to accommodate a large band, dancers and technicians. The sides of the *trio* are wall-to-wall speakers. The *trios* belong to carnival clubs called *blocos*. Members of the *blocos* wear a costume-tunic (*abadá*). Some members are true insiders in the organization, while others are consumers who change *blocos* each year, or tourists. The *bloco* occupies a mobile rectangle which is several hundred meters long and almost as wide as the street. At the center is the *trio*. The extremities of the rectangle are demarcated by a cord, held by a hundred or more workers (*cordeiros*) who advance up the street in pace with the *trio*. This rectangle provides a space for these members to 'play' (*brincar*, i.e., to dance and carouse) and have room to move around. In the intensely crowded space outside the rectangle which is left to the general public, hundreds of thousands of people also play carnival. The more daring accompany the progress of the *trio* and move past the throng on the sidewalk in the narrow space between it and the rope, or simply push through the throng. Standees have the alternate pleasure of viewing a succession of *trios* and their accompanying *blocos*. Both locals and tourists may purchase *abadás* or they may stay in the public domain and deal with the tussles in the throng. Bahian carnival is participatory and intensely physical, its contagious spirit mixing classes and nationalities in its various spaces. While often inept in the local Bahian dances and music, even the tourists are integrated in the action. Bahian carnival is both locally informed and assimilative of outsiders, both Afro-Bahian and cosmopolitan, suggesting the marriage of a deep sense of tradition and community with the innovative urgency of urban pop culture. The carnival is also a huge market for an informal economy of street vendors who sell drinks, food and knick-knacks, operating out of any space they can find in the crowd or between it and the buildings (Figure 1).

The three major, original carnivals of Brazil are those of Rio, Salvador and Recife (or Recife/Olinda, meaning the capital of Pernambuco state and its adjoining sister-city, the former capital). São Paulo has more recently developed a substantial carnival in imitation of Rio. The Rio carnival, at least in the form familiar to outsiders, is a spectacle held in the *sambódromo* – an arena which looks like a street but is in fact a completely controlled space for which one buys an admission ticket. The music of the Rio carnival is a special genre of samba, essentially restricted to carnival performance and rehearsals and stylistically stable. The Rio carnival is competitive in a very serious and rigid way. Its clubs – suggestively called *escolas-de-samba* or samba schools – must obey extraordinarily fine-tuned rules governing visuals, music, thematic interest and duration (not to exceed 65 minutes). The rules allow for gradual increases of extravagance of visual display in the costumes and floats, but judge dance-steps and percussion according to classical parameters, thus discouraging major change.

In contrast with Rio, the carnivals of Salvador and Recife are participatory events. Apart from the costumes, there is little difference between the members of a group and the bystanders. The groups are recreational associations which may or

Figure 1. Street vendors in the carnival. Photo by Piers Armstrong.

may not represent a specific community or have some constitutive ideology. Both carnivals oscillate between communitarian affirmation and socially indiscriminate recreation. In both, performing groups pass before the masses in real streets, causing persons from very different social classes to rub shoulders. Whereas the Bahian carnival genres change rapidly, Recife's are relatively stable. The Recife carnival presents the most eclectic range of traditional performance ensembles, to each of which correspond certain musical and dance genres, as well as types of costumes. While these are not governed by performance rules as in Rio, the Recife carnival maintains folkloric traditions very deliberately; this agenda is its distinguishing feature and determines the core of its content. Stylistic change (as in the case of the influence of foreign or interstate genres) is resisted by collective assent and/or city council ordinances. In summary, while Rio's official carnival means spectacle, control and competition, the carnivals of Salvador and Recife appropriate real streets as their space and promote participation over spectacle; these two carnivals then differ in the contrast between the Pernambucan preservationist spirit and the Bahian tendency to stylistic mutation (see below).

In terms of international comparisons, the New Orleans Mardi Gras is similar to the Recife carnival in its rich amateur spirit, its preservation of a range of costume-types and its general cultural conservatism. Perhaps the most interesting comparison for Bahian carnival is to that of Trinidad. In terms of music, Trinidad carnival presents analogous features to Bahia in that two of its core genres, calypso and steel-drumming, are celebrated in the carnival but also performed regularly outside it. However, Bahian industrial pop has infiltrated carnival, while in Trinidad the traditional genres remain the staple. While there are some similarities to the theatrical verve of Trinidadian carnival, with its individual characters such as Midnight Robber, these personages have generally been relegated to folklore in Bahia. The professional large-scale spectacles by theatrical designers such as Peter Minshall bear comparison to the work of Rio's *carnavalescos* (the artistic directors of the *escolas-de-samba*, who design the floats and supervise costumes). In Bahia costumes are rudimentary, and the vehicles are more mobile soundstages than floats for visual display. While carnival's strong role as a bridge between Trinidadians and the world might appear to echo Bahia, there are important differences. Insofar as there is a Brazilian equivalent to the transnational dimension of Trinidad carnival (replicated by émigrés in Brooklyn, London, Toronto etc.), the model is more Rio than Bahia, with *escola-de-samba* music adopted by expatriate Brazilian communities in the mini-carnival days of various North American metropolises. Trinidad carnival prompts a major but temporary influx of expatriates (many from America) who reconstitute cultural nationality. In Bahia, the more important links are between international tourists and local Bahians (discussed further below), and, as there is also an influx of interstate tourists, between Bahia and other Brazilian region, thus accentuating a sense of regional identity within the national whole. Finally, Bahian carnival is not competitive in the manner of Rio (for the carnival clubs), or Trinidad (for calypso rappers, steel drum bands, the staged presentations of large *mas* companies, and traditionally costumed individuals). While officially prizes are awarded to the best club in various categories, most carnival participants are indifferent to these distinctions. The real competition is between carnival clubs, in advance, as they sell their uniforms based on the prestige of their contracted music groups, and then during the event, as each group emerges with relatively greater or lesser kudos, which is translated in the band's *caché* – its set scale of fees for appearances at various sorts of events. Performances outside carnival similarly buoy or deflate that prestige and *caché*. In other words, whereas in Rio or Trinidad carnival is like a sports championship whose winner gains a glorious title registered in the history books, in Bahia carnival is simply the biggest gig for bands seeking to do the same work year-round.

In Bahia, the culture of carnival, which primarily means the production and consumption of music and dance, but also extends to related entrepreneurial opportunities, is spread over the whole calendar of annual and weekly festivities, both religious and secular, and has transformed Bahian society in terms both of its internal patterns of recreation and of its relations with the rest of Brazil and of the world. The Bahian carnival is not distinguished by unique musical practices, remarkable plastic forms or extraordinary license in informal interactions, so that its particular success cannot be attributed to the manifestation in carnival of some magical existential zone. On the contrary, what is distinctive is its seamless relation to local pop culture, a well-organized facilitation of that commodity which is persuasive to locals and outsiders alike and the permeability of its temporal and spatial boundaries with the everyday. Within the comparative repertoire of New World carnivals, the capacity for reinvention and transformation, the consistence with quotidian cultural consumption, and the centrality of carnival practices in the material economy together constitute a distinct Bahian carnivalesque system.

Afro-Bahian affirmation in carnival

In Bahia as elsewhere in the New World, one aspect of black culture has been to creative resistance to the psychological brutalizations of racism. Here, carnival should be taken in the general context of black use of official civic festivities as a platform for ludic dissent. Such events appropriate central public space, bring symbolic agendas to the foreground and permit public jubilation by the masses, affording a temporary ambient of exuberance and freedom in which an enthusiast ostensibly cheering the flag could really be cheering something else. Bahian historians focusing on black urban mobilization in the nineteenth and early twentieth century, such as Wlamyra Alburquerque (1999) and João José Reis (2002), show how Afro-Bahians took advantage of the vital margin of social license and symbolic ambivalence to project alternative agendas articulating their own cultural identity, one denied in everyday social intercourse (Figure 2).

Amongst the different civic festivities, carnival affords the most license, and this freedom is multidimensional. The aesthetic agenda embraces performative virtuosity, novelty and even strangeness. Each group determines its own annual themes, prompting a popular self-referentiality, whether in terms of class or ethnic heritage. In Bahia, this carnival license traditionally existed in a state of tension with the public management of events, both in terms of basic crowd control by the police and in the parade-design (decisions governing which icons appear when).

As with the everyday order, carnival is mostly managed by persons from within the white elite and is scripted in a way which celebrates the status quo. The town administration decides which famous people to invite as official carnival guests (announced and cheered by the crowd); the path of the parade through the city; how the available public space along the way (at the lateral edges of the parade course) is

Figure 2. Personal fantasy: an older woman enacting a scene as a fortune-teller. Photo by Piers Armstrong.

temporarily allotted to associations; where 'informal-economy' street vendors can be; the timing of each performing group; and the official annual theme of the carnival (not binding on individual groups). While carnival traditionally provided a space for black expression, it tended to be at the event's physical margins. The central performances of carnival – those in the most famous squares allowing the largest assemblages, at the times of day when the largest number of spectators are present – were inserted in the official carnival program by civic authorities, and were either directly sponsored by the state (for example, when dignitaries parade in vehicles), or were privately sponsored but aesthetically and ideologically mainstream manifestations (such as pop singers).

This pattern changed with the emergence of Afro-centric *blocos*, called *blocos afro*, in the late 1970s. The *blocos afro* drew their names from Yoruba terms used in the Afro-Brazilian religion of Candomblé, such as Ilê Ayê ('our house in this world') and Olodum ('God of Gods'). Ilê Ayê was formed in 1974, at the end of the worst phase of the dictatorship, when Portugal's African colonies achieved independence through armed struggle. Ilê's appearance in the downtown *avenida* was an aesthetic revolution with profound socio-ideological implications. In systematic contrast with the *blocos*, their costumes were based on African clothes; they used only percussion instruments; and their themes were Afro-centric. Inverting the practice of many of the more exclusive *blocos*, Ilê did not permit whites, or even mulattoes, to join. Ilê was based in Curuzu, an off-shoot of the populous and somewhat distant suburb, Liberdade, so called because it had emerged as a haven for free blacks in the nineteenth century. The annual themes of Ilê and subsequent *blocos afro* covered African countries (Kenya, Madagascar, Zimbabwe), Afro-diaspora countries (Cuba) and ancient history (notably black Egypt) (Figure 3).

Unlike the conventional *blocos*, the *blocos afro* had their own bands, made up of large numbers of volunteer percussionists who played for musical enjoyment and the recreational privilege that membership of the group afforded. The *blocos afro* held contests to select a small set of official songs on the year's theme, in open public performances known as *ensaios* ('rehearsals'). The *ensaios* afforded a legitimate venue for young people to gather and interact and provided opportunities to

Figure 3. A candidate for the title of Queen of Ilê Ayê dances as part of the competition. Photo by Piers Armstrong.

emerging composers and performers. Olodum's musical director, Neguinho do Samba, harnessed regional percussion traditions and imported musical styles to create *samba-reggae*. This rather martial rhythm and muscular percussive style is very distinctive and constituted a completely new sound in Brazilian music (Crook, 1993). The competitive structure of the *ensaios* spawned a crop of exceptionally good songs, both in words and melodies, and many *bloco afro* songs became local hits when scored for pop and commercially recorded around the time the *blocos afro* penetrated the mainstream (Moura, 1987; Guerreiro, 2000).

The Afro-centric ideological agenda of the *blocos afro* was notable in national terms for its dissent from the conventional top-down rhetoric of Brazilian civilization, which down-played racism. The dominant position in the intellectual establishment – notably that of the anthropologist and former Minister of Education, Darcy Ribeiro (1995/2000) – argues for a new amalgamated Brazilian ethnicity which supersedes the original, continentally-based contributing ethnicities (European, indigenous and African). In addition to government policy makers, most intellectuals have rejected black militancy as an inappropriate ethnic separatism. The mainstream narrative takes pride in Brazilian civil society as a remarkably harmonious cultural union, albeit beset by circumstantial problems (crime, poverty, ignorance) attributable to material want. This unifying, apologist approach has been celebrated many times in the presentations by *escolas-de-samba* in the carnival of Rio, whose themes often present the development of Brazilian or universal civilization in a miraculous light. In contrast, the policy statements of the Bahian *blocos afro* consistently stress racism as a fundamental social divider and set it against black pride.

Another dimension of the *blocos afro* has been activist intervention in everyday needs. Apart from the role of the percussion bands in developing discipline amongst the players, certain of the *blocos afro* have developed their own schools as an alternative to conventional state education. Here, black identity (*negritude*) is given a positive value which counters the stigma of racism. Complementing these historic events, the globalization of the Bahian cultural economy in the form of cultural tourism, principally by Europeans sympathetic to Afro-centric emancipation, validated the *blocos afro*'s significance and suggested an emblematic importance on the world stage.

Though constituting only a small proportion of the overall range of clubs, the *blocos afro* commanded such aesthetic prestige and social kudos as to become the symbolic center of Bahian carnival. The shift from the margin to the center and the role of the *blocos afro* as a sort of collective lyric hero are the hallmarks of the 're-Africanization' of the Bahian carnival (Risério, 1981; Agier, 2000).

Globalization and commodification

The success of Bahian carnival is partly based on a dialogue in which tourist enthusiasm complements an emerging local Afro-centric pride. Despite the very different perspectives contributing and resulting from this alliance of interests, there is a strong symbolic investment in the significance of the event as black. This is most evident in the coincidence of separate symbols in the recuperated central historic quarter of Salvador, called the Pelourinho. The name refers to a pillory column used to publically punish slaves, which once stood in the central square of the area. In the eighteenth century, many wealthy families built mansions on this square; the area also has many notable baroque churches, and the composite is considered one of the major congregations of baroque architecture in the Americas. By the twentieth century the area had deteriorated economically, and very poor families squatted in the mansions. Because of the centrality of the district, its demographic density, and the concentration of churches important for traditional cerimonial practices, it remained a meeting place for the Afro-Brazilian religious and creative communities, a cultural *agora*. The streets of the area were frequented through the mid-century by local Bohemians and later by tourists, mostly European. As the Church owned many buildings and gradually allowed more diverse lay community and foreign charitable organizations to use them, by the 1980s it had become an NGO-hub. With the consolidation of the industries of cultural tourism and carnival, the real estate in the area was restored and opened by the state to the initiatives of individual capitalists. The Pelourinho came to stand for a symbolic and spatial convergence of all these factors: black cultural continuity and political awareness; a cultural tourism marking the synergy between this local reality and world recognition; the recuperation of architectural heritage, its recycling into commercial infrastructure; and the commodification of Afro-Bahian culture.

The *blocos afro* thus transformed carnival performance, its symbolic politics and international perceptions of Bahia. Just as importantly, on the ground, the weekly *ensaio* system made the carnival subculture the staple of everyday recreation, while the performative ambient made for constant musical innovation, to a point where it colonized urban popular youth culture. At the same time, the thematic core of the songs was a distinctive ethnic identity. Tourism consecrated this symbolic system and transposed it into material infrastructure, thereby both empowering the progressive agenda and the commodifying the cultural heritage.

All these relations are symbiotic and they often seem politically indecipherable. The musical history itself may be the most reliable archive of meaning, and it illustrates how re-Africanization was eventually counterpointed by the inverse: the marginalization within the blocos afro themselves of the re-invented percussive sounds, with their overtone of ethnic dissidence, in favor of mainstream orchestration and deployment (i.e., guitars and synthesizers instead of raw percussion, on huge trucks instead of on foot), and ever more commercial themes. The relation whereby carnival fed popular culture was also inverted. Whereas in the 1980s the dominant *axé* sound fed from carnival into pop, by the late 1990s, the reverse occurred. A distinct Bahian style of *samba pagode*, drawn from the urban samba of everyday consumption, became hugely popular. While its performers are predominantly Afro-Brazilian, and its dance steps and rhythms build on the Afro-Brazilian heritage, it is scatalogical and skeptical rather than ideological in terms of both communitarian bonds and ethnic identity. As a result, the style was avoided by both consecrated individual *axé music* performers *and* the *blocos afro*. In 2003 the mass popularly of *pagode* was challenged by a particularly sensual slow dance for couples called *arrocha*, which developed in provincial towns outside Salvador and which seems likely to be a passing but sensational sub-chapter of Bahian pop (as *lambada* was in the mid-1980s). The current triumverate of *axé music*, *pagode* and *arrocha* may last into the next decade, but will no doubt succumb to new sounds. The arc from the ideological to the commercial which pertains within the history of *axé music* plays out on a larger scale if we include these newer genres.

Carnivalesque aspects in Bahian political power and cultural citizenship: the perpetual negotiability of race relations

Further scrutiny of local power relations suggests varied roles for sectors of the largely white local elite: as members of a political establishment which traditionally discouraged black *cidadania* and dissent; as capitalist entrepreneurs in the new tourist and consumer markets created through the cultural movement; and as professional musical artists, riding the growing commercial swell of the wave of re-Africanization. The sheer agility of the negotiation of increasingly Afro-centric cultural commodities, by both progressive and conservative elements of the largely white middle and upper classes, suggests that re-Africanization and its management were not unprecedented but rather inscribed in an ongoing pattern of mutual accommodation between elite and popular agendas. This points to creolization as a process of strategic negotiation by whites as well as blacks, in which subalterns overcome barriers and elite groups periodically re-draw distinctions. The metaphor of fusion (as in the term 'melting pot') is often legitimately applied to creolization as a cultural process but is less apt in relation to the political games obtaining in a Creole society, which, while effecting a real communion, are also more about managing distinctions within the ensemble than about merging.

Despite its fascinating ideological symbolism and aesthetic creativity, the 're-Africanization' should probably not be taken as an unique transformation, but rather as the most vivid manifestation to date of the capacity of Bahian carnival to generate new forms at its margins and then assimilate them into its mainstream. The stylistic dissidence of the re-Africanization phase has proved compatible with the reiteration of the core performative genre constituted by the *trio elétrico* which is mobilized by

Figure 4. Crowd enthusiasm at a competitive performance of drag queens which emulates Trinidadian carnival. Photo by Piers Armstrong.

conventional entrepreneurial interests. From the point of view of the *blocos afro*, however, their own assimilation of entrepreneurial strategies was a successful bottom-up penetration of the marketplace, reflecting partial reinvestment of the substantial symbolic capital accumulated during earlier years (Figure 4).

The centripetal effect of Bahian carnival on social mores has analogues in both the broader leisure industry and in Bahian politics. Many Bahian critics have pointed to the notion of *baianidade* – an ethos of personal well-being and affable social communion considered to be regionally characteristic and linked to carnival – as a disincentive to debate and due process, and also to cultural dissidence. In more concrete terms, the history of Bahian democracy should be cast in the context of Brazilian democratization after two decades of dictatorship from 1964 to 1985 (Butler, 1998; Pinho, 1998; Oliveira, 1999). The topic goes beyond our present scope, but several aspects resonate with processes described above in relation to carnival. On the one hand, Bahian politics is centralized – dominated by central discourses and persons – to a degree which is high even by Brazilian standards; on the other, as in the case of the *blocos afro*, this same effect does not simply mean the displacement of groups at the political margins, but also an ambiguous incorporation of the socio-economic and cultural margins. The tendency to bring together different social stations, obfuscating a persistent underlying hierarchy, is reminiscent of the social choreography of carnival. In the terms of this essay, as a sensibility and everyday practice, it is carnivalesque and also centripetal in two senses: vertically, in class terms; and horizontally, in cultural terms. Along the first axis, the code of informal camaraderie draws rich and poor together; along the second, in a society which is liberal but conventional, the gregarious impulse induces cultural others (i.e., not so much subalterns as 'deviants' – punks, gays and artists, for example) into an ecumenical communion with 'ordinary people'.

This centripetal tendency is, however, a two-edged sword, in that a restrictive repertoire of cultural figures – a pantheon of carnival icons – displaces less emblematic entities. The cultural tourism which constitutes a growing proportion of overall economic activity in the region around Salvador is predicated on the reliable performance of iconic features of Bahian culture and hence on an essentialized version of the cultural panorama. The most emblematic icon is the *baiana* – not a woman from Bahia state but rather a black woman who makes and sells in the street foods derived from Bahian cuisine and associated with Candomblé. The vendors are in fact obliged by municipal regulations to wear the distinctive colonial garb of white cotton lace used in Candomblé, regardless of their personal religious orientation. Since it pre-dates modern tourism and even modern carnival, the semantic slippage from the adjective denoting all women in the state to the nominalized adjective connoting a religious minority is evidence of a long-standing centripetal inclination in Bahian culture. *Capoeiristas* are another genre of iconic figures; like the *baianas*, they exercise a traditional art which pre-dates tourism and, while amenable to it, is independent from it. For their part, singers and dancers by and large conform to the stiflingly narrow stylistic confines of *axé music*. While eccentricity is welcome in any carnival, a carnival-centric culture can militate for a very superficial form of pluralism (Figure 5).

Performance is more purely responsive to image when created exclusively for it. Just such a performative niche has emerged in the form of fantasy *baianas* – i.e., women who are not food-vendors but who, like Carmen Miranda, wear costumes

Figure 5. *Tropicalista* pop stars, Gilberto Gil, Gal Costa and Caetano Veloso (from left to right), enact the arrival of the Portuguese discoverers of Brazil. Photo by Piers Armstrong.

derived from that tradition. These glamorously produced women stand around in the Pelourinho, apparently idle. They engage tourists in friendly conversation, encourage the latter to take pictures with them, then ask for payment for the privilege, pointing out that their cultural performance is achieved at considerable expense. The request for payment may be before or after the picture and is sufficiently informal as to precipitate negotiation, adding a certain dramatic verve to proceedings and making the tourists unwitting co-actors. Their outfits are heavily stylized but unique, just as their economic interest is highly individualized. Their performance radicalizes the process of essentialization because it is personally incorporated and restricted to informal social intercourse, rather than existing as the impersonal execution of an established art by a professional performer. The interactivity and fantasy of the incorporation recalls the traditional motif of carnival as encounter between masked and unmasked individuals in public space. The fantasy *baiana* thus works on tourist expectation by drawing on the carnival heritage as a cultural resource, and further cements carnival to the everyday. In terms of economic citizenship, the practice allows people without more conventional assets in the labor market to become players in the touristic center (Armstrong, 1999).

This informal cultural economy may appear to have little to do with politics. In fact, the Bahian state government is the chief financer and thematic arbiter of tourist advertising, and the main employer of fantasy *baianas*, and is thus directly involved in the promotion of cultural stereotypes. It also plays a very effective intermediary managerial role between the popular masses who perform traditional cultural activities and the outsiders who come to see them. The heavy police presence in the Pelourinho is another reminder of the state's constant proximity. Bahian tourist promotion draws on both traditional folkloric icons and markers from youth pop culture. One of the first images encountered at the airport is a large billboard of a smiling dreadlocked Afro-Bahian wearing the colors of Olodum, and bearing the paddles used by airport workers to direct incoming planes. The image thus forges a symbolic bridge between the Pelourinho, international tourism and a modern economic infrastructure. The centrality of carnival is constantly re-iterated in official slogans like '*Vem para a Bahia – é carnaval o ano todo!*' ('Come to Bahia – it's carnival all year round')

The representation of carnival is disproportionately Afro-centric, privileging *blocos afro* and *afoxés*, though they constitute a small proportion of the total number of carnival clubs, because they successfully appeal both to international tourists and the preponderantly Euro-Brazilian tourists from the Brazilian Southeast. The local white oligarchy, which completely dominates the party system, is effaced from the representational visage of Bahia. This Afro-centrism generally does not transcend the symbolic to penetrate the material sphere. The modern infrastructure implied by the airport billboard is almost entirely controlled and directed by Euro-Bahians, following North Atlantic models, authorities and capital. In fact, the Bahian government blatantly uses carnival to allieviate popular anger in the face of poor material infrastructure and inefficient deployment of resources due to corruption.

As in most regions of Brazil, Bahian democratic politics is heavily influenced by earlier long periods of authoritarianism and operates through networks of personal favors and alliances, often termed *clientelismo*, i.e., dependence on powerful patrons. *Clientelismo* tends not only to reflect but also to foster personalized power relations and position-taking on issues; most political parties in Brazil are based on complex strategic alliances rather than consistent ideological platforms This tends to promote vote-buying, circumventing democratic participation through public pressure groups, accountability checks and balances. From the general perspective of the Brazilian Southeast and Brasília, Bahia has particularly old-fashioned power relations, even in comparison to the neighboring state of Pernambuco, whose capital, Recife (the other large city of the Northeast) has undergone a more substantial democratic transition and presents greater political volatility.

The political leadership in Bahia was until recently incarnated by one man, Antonio Carlos Magalhães, referred to by his initials, ACM. In a regal allusion, his system of political domination was called *carlismo*; to some extent it has survived his passing (in July 2007). *Carlismo* presents an unparalleled continuity over almost forty years. ACM was one of the few politicians to have exercised major power in alliances with both the military dictatorship and then, in the democratic era, with the civilian politic order. He was mayor of Salvador, State Governor, Federal Minister and a federal senator. His protégés led city and state governments for most of the period, and his face was displayed with his candidates in their campaigns. ACM is widely believed to have had rebellious immediate family members killed, in addition to dissidents during the dictatorship period. He remained a central powerbroker in Federal politics after surviving major scandals, and, due to a stint as Federal Communications Minister when TV licenses were renegotiated, directly or indirectly controlled much of the Bahian media (Gomes, 2001). ACM was an iconic and frequent figure in civic and cultural events, from visits by foreign dignitaries to the carnival and concerts by prestigious artists. In short, befitting a carnival society, the play of power under ACM was akin to that of a court.

The administration of the carnival, including the granting of licenses needed to financially exploit some aspect of its cultural capital, is highly susceptible to the infiltration of *clientelismo*. The build-up to carnival is an extravaganza of information dissemination, and the event itself a smorgasbord of publicity opportunities for official and allied interests. Surprisingly, in relation to the *blocos afro*, ACM's group managed to dominate and even win the official alliance of the most important *blocos afro*, and to prevent the emergence of alliances between black political militants, the black cultural elite and the Left, despite their common agendas of denunciatory social rhetoric. Such control is exercised at the most basic level. The *blocos afro* and other activist groups depend greatly on the state for infra-structural support for everything from leases on buildings in the historic center to the channeling of international sponsorships for educational projects.

Earlier in this essay I argued that though Achille Mbembe's notion of a civic carnivalesque in post-colonial authoritarian African regimes should not be directly applied to Latin American emerging democracies and carnival societies, it nevertheless shows how a carnival aesthetic may permeate the everyday cultural order. The Bahian political scenario is very susceptible to such a reading. It is virtually uncontested in Bahian academic circles that most scholars refrained from open political criticism of the regime for diplomatic reasons at various points in their professional lives. There was also a general sense in Bahia – of lesser degree but analogous to that in Cuba (perchance, Fidel Castro, who would visit both Bahia and ACM, was born in the same year, 1927) – that some massive generational shift would come with the death of its seminal autocrat. It is still too soon to gauge any such shift. Partly as a result of self-muzzling by academics, there is a dearth of solid research on power relations and on whether the extraordinary career of one politician is paradigmatic or exceptional. It will be more telling if it proves paradigmatic.

It is tempting to read this personalized domination – call it clientelism, or just the 'court' – within a cultural logic of the carnivalesque. ACM, living or dead, casts a shadow on Bahian civic culture. Of course, this may just derive from this critic's personal memories: wandering around the Pelô and coming across a public banner hung by (or for) Ilê Ayê, congratulating ACM on the occasion of his birthday; or, in the middle of a concert by Caetano Veloso, Bahian pop star, seminal tropicalist and former political exile, seeing the spotlight suddenly swing to the back of the arena, to settle for a moment on a Caesar-like ACM. Still, the figure of ACM illustrates, in Technicolor, the centripetal influence of carnival. The proclivity to personification and theatricality are culturally inculcated in carnival societies – whether through mental habits developed by performance, or, rather, because such features, and carnival itself, are symptomatic of a deeper stratum of psycho-social values – liberating, atavistic and still mysterious.

Conclusion

We have to be cautious not to over-interpret the pattern of influences between carnival and the everyday. In the Bahian case, the social *carnivalesque* deriving from the cultural traditional has historical roots pre-dating the contemporary carnivalization and even carnival itself. As a social phenomenon the everyday carnivalesque may prompt carnival performance rather than the reverse. Without resolving primacy of origin or causality, what may reasonably be asserted is the mutually influential relation between carnival and everyday life. This relation presents a socially centripetal, or 'centralizing', characteristic which manifests itself in different domains.

The execution of carnival is predicated on a colletive performative learning, and its particular spirit of play implies both a collective cultural initiation and an authentic cultural union. In addition to its ludic virtuosity, a range of material and other pragmatic agendas and interests are built into this socio-aesthetic logic, and these are susceptible to more objective analysis as social trajectories. In this regard, the Bahia case study, like others, suggests the interaction of top-down and bottom-up strategies, so that the centripetal tendency of carnival should not be read simply as a reiteration of established cultural order, but also as an opportunity, albeit ambiguous, for marginalized individuals or groups to cast themselves into the middle of the social arena – whether for reasons of self-affirmation or for money.

If Bahian carnival is representative of its cousins, carnival in the Americas is not merely invested with black traditions in a circumstantial, practical way. Rather, a theatrical black presence is a consecrated element, and diaspora carnival is a sublimation of the creolization of blacks and whites as a highly complex process of mutual accommodation, involving material issues and rational strategies, on the one hand, and, on the other, emotive relations and mythical narratives transposed into the popular performative vocabulary.

Given the demographic variation in the range of New World carnival societies, the racio-social dynamics and underlying mythologies may vary widely. The Bahian carnival is distinctive in its stylistic volatility, its relation to youth culture, and its bi-polar embrace of Afro-centric aesthetic discourses and Euro-centric material authority.

References

Aching, G. (2002). *Masking and power: Carnival and popular culture in the Caribbean.* Minneapolis: University of Minnesota Press.

Agier, M. (2000). *Anthropologie du carnaval: La ville, la fête et l'Afrique à Bahia.* Paris: IRD; Marseilles: Parenthèses.

Alburquerque, W.R. de. (1999). *Algazarra nas ruas: Comemorações da independência na Bahia (1889–1923).* Campinas: Unicamp CECULT-FAPESP.

Armstrong, P. (1999). The aesthetic escape hatch: The mutations of baianidade under the signs of globalization and re-Africanization. *Journal of Iberian and Latin American Studies* [AILASA, Australasia], *5*(2), 65–98.

Bakhtin, M. (1984). *Rabelais and his world.* (H. Iswolsky, Trans.). Bloomington: Indiana University Press.

Butler, K.D. (1998). *Jinga baiana*: The politics of race, class, and power in Salvador, Bahia. In H. Kraay (Ed.), *Afro-Brazilian culture and politics: Bahia, 1790–1990s* (pp. 158–175). Armonk, NY: M.E. Sharpe.

Cavalcanti, M.L.V. de C. (1999). *O Rito e o tempo: Ensaios sobre o carnaval.* Rio de Janeiro: Civilização Brasileira.

Crook, L. (1993). Black consciousness, samba-reggae, and the re-africanization of Bahian carnival music in Brazil. *World of Music*, *35*(2), 90–108.

DaMatta, R. (1991). *Carnivals, rogues, and heroes: An interpretation of the Brazilian dilemma.* (J. Drury, Trans.). Notre Dame: University of Notre Dame Press. (Original work published 1978)

Dantas, M. (1994). *Olodum: De bloco afro a holding cultural.* Salvador: Grupo Cultural Olodum/Fundação Casa de Jorge Amado.

Falassi, A. (1987). *Time out of time: Essays on the festival.* Albuquerque: University of New Mexico Press.

Fischer, T. (Ed.). (1996). *O carnaval baiano. Negócios e oportunidades.* Brasilia: SEBRAE.
Gomes, J.C.T. (2001). *Memórias das trevas. Uma devassa na vida de Antônio Carlos Magalhães.* São Paulo: Geração.
Guerreiro, G. (2000). *A trama dos tambores. A música afro-pop de Salvador.* São Paulo: Editora 34.
Hirschkop, K. (1999). *Mikhail Bakhtin: An aesthetic for democracy.* Oxford: Oxford University Press.
Jarry, A. (2003). *Ubu roi.* [Stageplay, 1896]. (B. Keith & G. Legman, Trans.). New York: Dover.
Karlstrom, M. (2003). On the aesthetics and dialogics of power in the postcolony. *Africa, 3*(1), 57–76.
Mbembe, A. (1992). Provisional notes on the postcolony. *Africa, 62*(1), 3–37.
Moura, M. (1987). Faraó, um poder musical. *Cadernos do CEAS. Revista do Centro de Estudos e Ação Social da UFBA, 112*, 10–29.
Oliveira, C.L.P. (1999). Struggling for a place: Race, gender, and class in political elections in Brazil. In R. Reichmann (Ed.), *Contemporary Brazil: From indifference to inequality* (pp. 167–177). University Park, PA: Pennsylvania State University Press.
Pinho, O.S. de A. (1998). A Bahia no fundamental: notas para uma interpretação do discurso ideológico da baianidade. *Revista Brasileira de Ciências Sociais, 13*(36), 109–120.
Queiroz, M.I.P. de (Ed.). (1992). *Carnaval brasileiro – o vivido e o mito.* São Paulo, Brasiliense.
Reis, J.J. (2002). Tambores e temores: A festa negra na Bahia da primeira metade do século XIX. In M.C.P. Cunha (Ed.), *Carnavais e outras f(r)estas: Ensaios de história social da cultura* (pp. 101–156). Campinas, São Paulo: Unicamp.
Ribeiro, D. (2000) *The Brazilian people: The formation and meaning of Brazil.* (G. Rabassa, Trans.). Gainesville: University Press of Florida. (Original work published 1995)
Risério, A. (1981). *Carnaval ijexá: Notas sobre afoxés e blocos do novo carnaval afro-baiano.* Salvador: Corrupio.
Rodrigues, A.M. (1984). *Samba negro, espoliação branca.* São Paulo: Hucitec.
Schechner, R. (2004). Carnival (theory) after Bakhtin. In M.C. Riggio (Ed.), *Carnival: Culture in action: The Trinidad experience* (pp. 3–11). New York: Routledge.
Sodré, M. (1999). *Claros e escuros. Identidade, povo e mídia no Brasil.* Petrópolis: Vozes.
Turner, V. (1969). *The ritual process. Structure and anti-structure.* Chicago: Aldine.
Turner, V. (1987). Carnival, ritual and play in Rio de Janeiro. In A. Falassi (Ed.), *Time out of time: Essays on the festival* (pp. 74–90). Albuquerque: University of New Mexico Press.

'The return of Crazy Mother': The cultural politics of carnival in 1930s Dijon

Philip Whalen

Department of History, Coastal Carolina University, Conway, USA

The political elite in inter-war France sought innovative ways to reconcile modernization agendas with existing regional cultural traditions. Regional folkloric traditions and ludic festivals provided refuge from maligned national narratives. Burgundians, notably, drew upon carnival practices and motifs to solicit broad participation in a state-sanctioned project of regional modernization. Although organized by ostensibly opposed political, social, commercial, and cultural interests, Dijon's Mère-Folle carnival and subsequent carnaval parades accomplished the same goals. These included creating a new spirit of civic pride and unity, providing a time and space for overlapping and converging transformations to coexist, and stabilizing cultural practices deemed traditional.

Dijon's 1935 carnival festival and subsequent, annual carnaval parades provide a fascinating example of how diverse political leaders, commercial interests, cultural intermediaries, and ordinary citizens collaborated to produce spectacles calculated to promote civic unity and collective gaiety. Following Peter Stallybrass and Allon White's observation that questions concerning 'the politics of Carnival cannot be resolved outside of a close historical examination of particular conjunctures' (Stallybrass & White, 1986, p. 16), my chief concern is to ask how emergent, preferred, and/or alternative readings of carnival might be produced, circulated, and consumed by variously invested social actors on the eve of and during the Popular Front in France, from 1935 to 1939. Aware that the relation between social practices and narrative possibilities (Jameson, 1981) involves an analysis of 'the spectacle' as 'a social relationship between people that is mediated by images' (Debord, 1995, p. 12; Schwartz, 1998), I propose to examine Dijon's modern carnival as a social text mirroring practical, political, and theoretical priorities by examining whose ideological, cultural and political interests were served by the use of carnival themes and motifs to structure civic events.

This investigation contributes to existing scholarship concerning Burgundy's inter-war regional cultural project (Poirrier, 1991). That is, how interested parties used folkloric spectacles to link regional traditions, popular culture, mass politics, and consumer practices into local identity projects susceptible to economic exploitation (Laferté, 2002, 2006; Moentmann, 2003; Bazin, 1997; Whalen, 2008).

As in the development of gastronomic tourism, Dijon's festivals illustrate the unexpected degree to which modern consumer practices and traditional cultural agendas were entangled, interdependent and politicized (10 M 121 and SM 12171 'Foire gastronomique' and SM 299 'Foires et marchés', Archives départementales de la; Guilard, 1999; Whalen, 2008). In support of this interpretation, I first discuss some theoretical questions concerning the significance of public spectacles and then relate them to carnival in both seventeenth-century and twentieth-century Dijon in order to recognize how carnivalesque art forms can affirm, contest and/or renegotiate social, political and gender identities.

Revived traditions and 'applied folklore'

Inter-war France turned to 'traditional' motifs and repertoires to address divisive party politics, collective social disenchantment and discredited national narratives (Peer, 1998; Weber, 2000; Golan, 1995; Whalen, 2007a). Political leaders and provincial cognoscenti collaborated to orchestrate local festivals that would simultaneously embrace modern aspirations, retain conservative social values, and generate sentiments of civic solidarity through the pageantry and pedagogy of 'applied folklore' (Carcano, 1997; Shuldiner, 1998; Payne, 1998; Demossier, 1999; Weber, 2000; Agulhon, 2003; Laferté, 2002 and 2005; and Whalen, 2007a). Burgundy's festival culture illustrates how traditional symbolic practices retained purchase on the cultural, economic and political imaginations of the twentieth-century French (Laferté, 2006; Whalen, 2007b).

Dijon's civic leaders believed that 'a people that play are a happy people' (Thomas, 1937, p. 6). Dijon's previous mayor, Gaston Gérard, had organized municipal fairs and festivals in order to 'allow everyone to escape reality, chase away grey thoughts ... combat monotony ... [and] find happiness within themselves' (Gérard, 1935, p. 3; Morhain, 2009). It was Gérard who had created the Burgundian Association for the Maintenance and Revival of Popular Feasts and Traditions in 1928 to encourage and support 'manifestations of popular traditions...within a mercantile perspective' (Vachon & Gerley, 1992, pp. 8–9). The Association organized what it considered to be an 'authentic' carnival 'reconstructed according to archival sources'– complete with puppets, parades, floats, music, and general merriment – in 1935 ('Quand ressuscite la Mère-Folle à sa bonne ville Dijon', *Le Progrès de la Côte-d'Or* 25 March 1935, pp. 1–2; Gérard, 1935). Despite limited archival evidence, newspaper accounts and a series of postcards reveal how traditional motifs and symbols were (re)coded to promote contemporary economic interests asnd allowed political differences to temporarily coexist in a shared civic space (Roupnel, 1922, pp. 153–154; Vachon & Gerley, 1992, pp. 1–9; Poirrier, 1994, pp. 377–389).

Carnival has rightly been defined as 'the example *par excellence* of the festival as a context for images and texts' (Burke, 1978, pp. 182–185). Usurprisingly, the cover of Dijon's 1935 carnival 'Official Program' reveals a fascinating juxtaposition of figurative signs, historical markers and regional cultural identifications (see Figure 1). At the top, above a carnival stage, an unfurled banner announces the 'return of Mère-Folle' and 'great pleasures'. Just below, against a silhouetted skyline of Dijon's downtown – in which architecturally noteworthy *ancien régime* buildings are recognizable – three figures revel on a rustic street stage. The central and largest character is dressed in jesters' clothing and holds a glass (presumably of wine) above

Figure 1. Marianne with jester's cap revelling in Dijon's 1935 Carnival. Cover of 1935 Mère-Folle 'Official Program'. Permission from Archives municipals de la Ville de Dijon.

the skyline's church steeples and the Ducal Palace's tower. He is seated on a barrel labeled 'Burgundy', which rests on a regional heraldic crest. He both symbolizes Burgundy's modern wine interests and recalls the alcoholic indulgence associated with early-modern carnivals. A startlingly modern female dancer whose leg kicks above several bottles of wine and glasses juxtaposes the foregoing Renaissance fair and carnival referents. Wearing tights (adorned with several bells) and striking a Folie-Bergères pose, she wears a jester's cap instead of the Phrygian-style liberty cap that would ordinarily mark her as Marianne – the idealized embodiment of revolutionary, republican, and therefore modern France (Burke, 1978, p. 191). Her arm, holding a tambourine on which the date 1935 is written, reaches across to the right side of the stage to the third character, a second jester or fool. He is blowing a horn – symbol of abundance traditionally associated with the Abbot of Conards (meaning both fools and 'whipped' men) and/or leveling, 'whether of food, property, status, wealth, or authority' – with one arm and waving a pair of balloons (which symbolize pigs' bladders, traditionally filled with urine and used to spray crowds in early-modern carnivals) on which the year 1630 is indicated (Scott, 1990, p. 167). In exchange, presumably for subsidizing the printing of the brochure, the bottom of the cover provides an advertisement for the Bruant Pharmacy. This promotional blurb riffs the French national 'Liberty, Equality, and Fraternity' slogan with one that underscores the qualities that constitute good wine, commerce, and tourism – 'Care, Quality, and Value'. In keeping with this agenda of commercial quality and profit, the handsomely printed program was – it is indicated – sold for one franc.

Spectacle as civic agenda in Dijon

While the idea of selling heritage is familiar, the actual mechanisms used to market and consume heritage remain problematic and difficult to locate. Pertinent to this discussion, Richard Dyer argues that 'abundance becomes consumerism, energy and intensity, personal freedom and individualism, and transparency and freedom of speech' in popular spectacles (Dyer, 1981, p. 185). Adopting the perspective that ideological investments may be (re)directed through performative practices such as play, consumption, tourism, and audience participation, the following examines the cultural logic and dynamic of Dijon's 1930s carnival and carnavals to better understand how regional festival culture engaged contemporary popular and ideological investments (Stallybrass & White, 1986, p. 16). The analysis of complex politico-cultural projects such as Dijon's Mère-Folle requires a creative interpretive strategy to understand how political tensions and cultural contradictions cohabited public stages. The simultaneous production of multiple experiences is a hallmark of carnival. By momentarily (re)framing shared public space, carnival can also (re)define the very experiences it stages.

The examination of how civic spectacles promoted consumer democracy in late nineteenth and early twentieth-century France necessarily raises the question of what role(s) participants imagined themselves to be performing (Rearick, 1985, 1997). Spect-actors, it has been suggested, find the momentary suspension of quotidian practices opportune for more than temporary release from stable social order and previous attachments. They produce a vision of alternative societal relations and structures that can both widen and confirm an array of behavioural options (Davis, 1975, p. 131). Attempts to generalize about the motives and behaviors of French consumers have ontologized a variety of socio-cultural types – from *flâneur* to *balladur* and from *badaud* to the 'tourist' that, while recognizing the modern consumer as a 'produced subject' or 'practical fiction', do not, unfortunately, explain the level or nature of ironic distancing that must have been necessary for adults attending Burgundian festivals, such as Dijon's 1935 carnival, to prance about the streets as though they were in attendance at the Court of the seventeenth-century Dukes of Burgundy (Nicolaÿ, 1902, pp. 1–2; Shaya, 2004, pp. 43–45; Nicole, 1933;

and 'L'inauguration de la XIième Foire Gastronomique', 1931, p. 2). Indeed, recognizing that heritage-making is a 'dynamic process of creation in which a multiplicity of pasts jostle for the present purpose of being sanctified as heritage', Jane Jacobs argues that it may no longer be accurate to assume that consumers or tourists seek 'authentic experiences' (Jacobs, 1996, p. 35).

Theoretical elaborations concerning the nature of 'spectacle' and the 'spectacular' (Debord, 1995, p. 151) help situate Dijon's carnival as a theatrical use of municipal space calculated to prompt a state of enhanced sociability and civic happiness not otherwise attributable to or synonymous with the contemporary political agendas. Frederic Jameson's assertion that 'the image is the final form of commodity reification' leads me to ask whether the construction of regionally defined patterns of consumption and civic participation in inter-war Burgundy reveal the operations of sophisticated cultural stratagems more often associated with post-war, consumer culture (Jameson, 1981, p. 276; Featherstone, 1991). While the ideological and practical intentions of the cultural intermediaries responsible for staging these spectacles are available through their own pronouncements, the subjective beliefs and social imperatives of the everyday citizens cast in the dual role of spect-actors that animated these events are notoriously difficult to discover. Fred Inglis' observation that 'ideologies maintain their constituencies, especially in their ritual and symbolic form because they are sufficiently general and vague to allow different groups and classes to find in them what suits them, and makeshift a social identity out of that' leads me to investigate the cultural links between modern consumer marketing practices to the development of regional political culture in inter-war France through the example of Dijon's Mère-Folle carnival and carnaval parades (Inglis, 1988, p. 108). Building on Debord's idea that the spectacle functions to separate the authentic from its illusion, one could equally follow John Urry's constructivism and conclude that 'post-tourists' simply find delight and satisfaction in the trappings of 'staged authenticity' (Debord, 1995; Urry, 1995, p. 140).

According to the circular hermeneutic in which audience/participants contribute unique personal itineraries to collective performances which in turn structure their experiences, Dijon's carnival (as polyvalent metaphor and performed ritual) brought people with diverse aesthetic interests, different ideological agendas and unique personal itineraries together for the purposes of inducing civic unity through festivity. The film director Marcel Carné, one may remember, used this stratagem to open and conclude his celebrated *Les Enfants du Paradis* (Children of Paradise, 1945). Staged during the turbulent July Monarchy and filmed during the Nazi occupation of France, the film's main characters' individual itineraries both emerge and disappear into an all-encompassing crowd of masked carnival revelers who represent *the human condition*. Similarly, a few years earlier in, Dijon's historic theatre troupe, the *Infanterie dijonnaise*, was revived to provide popular entertainment, promote commerce, and punctuate Dijon's cultural traditions. 'Five days of festival' were intended, according to the Committee for Festivals, to have important commercial repercussions, sustain the city's [tourist] reputation, confirm conventional gender roles, and at the same time distract all the classes of society (1I1/ 164 'Fêtes 1935 et 1936', letter dated 29 January 1935. Archives Municipales de Dijon).

Carnival in seventeenth-century Dijon

> Misrule can have its own rigor and can also decipher king and state. (Davis, 1975, p. 97)

Originally associated with Corpus Christi and Lenten celebrations related to Christian beliefs concerning 'the triumph of life over death and the sacrifice that ensures the redemption of human souls', the origins of Dijon's Mère-Folle carnival were linked to a religious fraternity that annually celebrated the Feast of the Innocents (Fête des Fous) in Dijon's Saint Chapelle during the fourteenth century. The Feast of the Innocents was recognized in Philip the Good's ducal decree of 1454 in which he 'desired, authorized, [and] ordered ... joyous innocents to dance unmolested' in public and sacred spaces (Martin, 2001, p. 7). As recorded in the registers of the Saint-Philibert church, the event acquired a secular character through close links to urban guilds centered in the Poissonnerie neighborhood and whose members tended vineyards outside the city's walls (Martin, 2001, p. 7). From these festive and pantagruelesque origins 'the Mère-Folle Society along with the *Infanterie dijonnaise*' notes historian Gaston Roupnel, 'created numerous theatricals no less inspired by Burgundian verve than its more refined and already French literature' (Roupnel, n.d.).

Recognizing that '[n]o Carnival was exactly like any other', Peter Burke provides the following composite of carnival as:

> a huge play in which the main streets and squares became stages, the city became a theatre without walls and the inhabitants, the actors and spectators, observing the scenes from their balconies. In fact there was no sharp distinction between actors and spectators, since the ladies on their balconies might throw eggs at the crowd below, and the maskers were often licensed to burst into private houses. The action of this gigantic play was set in more or less formally structured events. The less formally structured events went on intermittently for the whole Carnival season and were spread over the whole town. People sang and danced in the streets. People wore masks, some with long noses, or entire fancy dress. Men dressed as women, women as men; other popular costumes were those of clerics, devils, fools, wild men and wild animals. (Burke, 1978, pp. 182–183)

Looking to historical antecedents, Peter Burke reminds us, 'Carnival was the greatest popular festival of the year, and a privileged time when what oft was thought could for once be expressed with relative impunity' (p. 182). In Dijon, '[C]arnivalesque parades', led by a crossed-dressed fraternity captain (hence the word *travesti*) named Mère-Folle enacted 'urban charivaris' and pantomimed summary justice through mock trials, symbolic sieges, noisy affronts, demeaning *ripostes*, and farcical sermons (Davis, 1975, pp. 117, 99; Moriss, 1935, pp. 179–189; Scott, 1990, p. 157). Mikhail Bahktin reminds us that the 'popular-festive atmosphere' surrounding the winegrowers' annual grape harvest (*vendange*) permeates the second part of Rabelais' *Gargantua*, which describes a conflict between vintners and bakers that escalated into an all-out war (with attacks on Friar John's abbey) and culminated in a final and symbolic communion:

> [t]he figure of propitious time symbolizes in folklore the end of evil days and the advent of general peace. For this reason Rabelais develops a popular utopian theme: the triumph of peaceful labor and abundance over war and destruction. This is the

fundamental theme of this entire episode of the Picrochole war. (Bakhtin, 1984a, p. 228)

In another conflict, on 14 August 1482, the marchers of Dijon's *Infanterie dijonnaise* paid homage – *charivari* style – to the city's Mayor, the monks of several abbeys, and the rich bourgeois in their urban residences ('Le Retour de la Mère-Folle', 1935, p. 1). Royal governors, Masters of the Waters and Woods, and the Justices of Dijon's *Parlement* of Dijon were also targeted on other occasions. Natalie Zemon Davis notes that Dijon's Mère-Folle troupe 'ridiculed the king's Grand Master of Streams and Forests in Burgundy not only for beating his wife . . . but also for devastating – for his own profit – the forests he was supposed to protect'. This burlesque could contribute to an atmosphere fraught with existing conflicts and potential ruptures. 'Not surprisingly', Davis adds, 'the charivari and Carnival license to deride could also be turned against the political authorities' (Davis, 1975, pp. 117–118). Emmanuel Le Roy Ladurie offers the example of Mardis Gras in the southern French town of Romans in 1580 that devolved into public rioting along pre-existing socio-political divisions, pitting the town's artisans and workers against better-off landlords and rich merchants (Le Roy Ladurie, 1979). Despite these examples of carnival leading into outright rebellion (Basel, 1376; Bern 1513; Northern England, 1536; Dijon, 1630; Normandy, 1639; Palermo, 1648; the Vivarais, 1670; Madrid, 1766; Paris, 1789), Stallybrass and White maintain, 'there is no *a priori* revolutionary vector to Carnival and transgression' (Stallybrass & White, 1986, p. 16). The historical record does show that early-modern revolts – in any form – rarely

Figure 2. A seventeenth-century Mère-Folle standard (artist unknown). Permission by *Pays de Bourgogne.*

succeeded in effecting more than momentary change (Bercé, 1994, pp. 80–81; Reid, 2001; see Figure 2).

Mack Holt, among others, challenges interpretations that reduce social relations to a continuum of popular and elite antinomies as, for example, in Robert Muchembled's magisterial *Popular and Elite Culture in France, 1400–1750.* Holt finds evidence of workable compromises and solutions in Dijon during the early-modern period (Muchembled, 1985). He identifies the popular election or royal selection of the Mayor of Dijon as illustrating moments in which tensions, disorder, and divisions were scripted and resolved through popular food and wine rituals (Holt, 2000, pp. 21–29). Indeed, the legitimacy of rituals surrounding elections and eventually recognized when appealed by the Parlement (High Court) of Burgundy following an earlier attempt to prohibit the 'giving of either food or drink to the said wine-growers and aforementioned men on the day of said election, under penalty of an arbitrary fine' (p. 23). However-much wine growers' and artisans' rambunctious 'outbreak of noise which interrupted the social decorum' might be perceived as 'seditious', their involvement rewrote previous electoral practices to expand popular participation and arguably to instantiate local ties and traditions of 'moral economy'. The happy result of local practices could, however, be trumped by royal authority. Although Henri IV acquiesced to regional interests by selecting the Mayor of Dijon that had received a majority of local support, Louis XIII, instead, sought to exclude 'the inhabitants who were the poorest and among the lowest condition' by imposing poll taxes and, on the occasion of his visit to Dijon in 1629, refusing to guarantee the city's rights and privileges (Holt, 2000, p. 29).

Safety-valve interpretations of social performance which argue that popular spectacles manage social, cultural, and political contradictions in order to 'make then seem to disappear' rest on the assumption that cultural meanings are both created and manipulated by their producers (Dyer, 1981, p. 185). The dazzling masked balls of Napoleon III's 'Carnival empire', for instance, have been interpreted as offering participants the opportunity to enter 'worlds of fantasy in which the extravagance of the occasions distract the eye from disagreeable realities' (Horne, 2004, p. 241). Against a view of mass culture as bound by predictable practices; constituted by 'passive cultural dupes ... caught up in the static, totalistic, and inescapable web of ideology'; and frequently resting on 'uniform and totalizing versions of the psychoanalytic subject', John Fiske finds popular audiences not limited to the logic of the false choice but rather 'capable of producing multiple and preferred interpretations' (Horkheimer & Adorno, 1995, pp. 120–167; Mayne, 1994, p. 157; Fiske, 1989, p. 9). Indeed cultural studies argue that it is precisely at the level of consumption that meanings are simultaneously produced, altered, and managed (Baker, 2002, p. 167).

John Fiske interprets the transgressive potential of carnival(esque) behavior as having, 'on some occasions, in some contexts', the potential to strengthen social order while, on other occasions and particularly in times of social tension, the possibility to be much more disruptive (Fiske, 1989, p. 100). Unable to resolve the agency/structure assumptions that support such interpretations, Stallybrass and White caution that '[i]t would be wrong to associate the exhilarating sense of freedom which transgression affords with any necessary or automatic political progressiveness'. Rather, they suggest that the semiotic production of literary meanings, social manners, and body decorum construct the world in terms of an elite 'self-identity' by subliminally encoding the 'historical being' (Stallybrass & White, 1986, pp. 201–202).

If not outright subversion, carnivals offer the possibility, at the very least, of having created, through inversion, a space for renegotiating and even resisting culture. While Stallybrass and White read carnivals as mere cooptation through licensed complicity insofar as official dominant culture is not challenged, Johns Fiske argues, instead, that:

> progressive practices are panaceas allowed by the system to be flexible and to contain points of opposition within it. By allowing [these], progressive practices actually strengthen that to which they are opposed, and thus delay the radical change that is the only one that can bring about a genuine improvement in social conditions. (Fiske, 1989, p. 192)

But what if socially disruptive play, excess, and even riot can also produce 'a demystifying potential' that resists policing (Stallybrass & White, 1986, pp. 18–19, 192)? Synthesizing methodological and topical approaches to carnival theory, Denis-Constant Martin reframes Bakhtin's literary study of medieval festival practices (as they figured in Rabelais' non-theatrical works) to underscore the significance of ambivalence rather than inversion (Martin, 2001, p. 5). In such terms, it remains possible (at the social rather than structural level) to recognize the spaces or moments in which official culture is made to 'back off' and therefore concedes terrain to a savvy, self-conscious, oppositional spect-actor culture.

In addition to negotiating political dissent, social conflict, religious tensions, or political aspirations, Dijon's early seventeenth-century carnival occasionally devolved into what Counter-Reformation authorities perceived to be seditious disturbances. Recognizing that the antics of Mère-Folle (Crazy Mother), her husband Bontemps (Father Good Times) and Le Peuple (The People) had genuinely bridged 'the mentality of the *parlementaires* (nobility of the robe) and the urban crowd's passionate, malicious and credulous' disposition, the historian Gaston Roupnel further noted its mid-century demise as reflecting a broadening cultural divergence related to urban socio-economic polarization. The themes addressed by Dijon's Infanterie Dijonnaise became too provocative for Dijon's authorities (Roupnel, 1922: 150–153). Because of its devolution into bloody rioting (associated with vintners' tax grievances and Louis XIII's chilly response to the City's warm overtures) Cardinal Richelieu suppressed the Mère-Folle in 1630 (Patouillet, 1971). It briefly revived 'in all of its luster' in 1638 when 400 revelers marched to celebrate the birth of the Dauphin, the future Louis XIV (Chabeuf, 1897, p. 29). Subsequent revivals, however, were limited and increasingly infrequent. They celebrated a royal visit in 1650 and punctuated the accomplishments and visits of Burgundy's own Condé family (1660).

Carnival, it has been argued, produces a middle space constituted by both structural antinomies and heteroglossic uncertainties which characterize the shared, multivalent experience of a spectacular reality that is ambivalently dependent on dominant order and whose expressions range from begrudging accommodations, symbolic inversions, oppositional farts, aesthetic evasions, and/or oblique ripostes (Davis, 1973; Bakhtin, 1984b; Kellner, 2003, p. 29). Insofar as carnival can provide different, overlapping, or competing agendas the opportunity to coexist in time and space, one cannot be quite certain, without first person testimonials, about how others have understood them. Were they viewed simply as diversionary aesthetic spectacles, pseudo-historical charades, political window-dressing or cultural liturgy come-down from superiors? When or where might the use of fantasy or utopian rhetoric be used to produce the effect wherein 'selfhood is dissolved into the intensity of fantastic experience in order to make the impossible momentarily real' (Inglis, 1988, p. 105)?

Crazy Mother's final transgression and execution

As grotesque masks flagrantly represent the jockeying embodiments of cultural contradictions, intensified symbolic dissonances, and displaced ideological distortions (Aching, 2002), carnival stages a dizzying range of theatrical possibilities for grotesques to narrate– in what Bakhtin calls 'a concretely sensuous, half-real and half-play-acted form'– preferred civic allegories and collective dreams (Bakhtin, 1984a, p. 123). Carnival spect-actors (re)define themselves and leave previous attachments by playing new roles – however contradictorily and/or unpredictably scripted – as consumers within manufactured, pre-fabricated contexts or simulated 'pseudo-events' in order to playfully (re)produce and negotiate heterogeneous meaning(s) (Williams, 1994, pp. 157, 159). Against structuralist interpretations that privilege popular opposition to established hierarchies, Stallybrass and White provide a deconstructionist interpretations of carnival that underscore ambivalence, ambiguity and indeterminacy. They argue that the logic of identity-formation involves 'distinctive associations and [the] switching between location, class and the body, [as] these are not imposed *upon* subject-identity from the outside, are [rather] the core terms of an exchange network, an economy of signs' that include carnival's 'imaginary repertoire of transgressive desire, economic tensions, and political contradictions' (Stallybrass & White, 1986, p. 25). Such operations are clearly pertinent to the role of women in Dijon's 1935 carnival.

Dijon's 1935 Mère-Folle revival was both a popular and administrative success. 'What a return', exclaimed a local booster: 'a parade of floats, a torchlight ceremony, an evening bal, a pitched battle between gigantic paper dolls, a banquet for 1000 persons, and, finally, the mock trial, confession, testament, and execution (by fire) of Mère-Folle's effigy' (Vachon & Gerley, 1992, pp. 8–9; see Figure 3)· Dijon's historic centre was festooned and illuminated with original decorations that produced, according to one journalist, 'a unique cachet of attractiveness to all the festivities' (*Bien Public*, 22 March 1935, p. 3). The event invoked obscure activities and nostalgic memories within the realm of modern mass culture. Robert Houdart temporarily transformed Dijon's Hotel de Ville's parking lot into a Ducal courtyard and marketplace and simultaneously into an object of consumption, a space for social performance, and a vehicle for historicized representation (Gérard, 1959, p. 40). Newspaper accounts roundly agreed (sententious and censorious interpretations aside) the event was attended by representative cross-sections of Dijon's social groups. The photographic evidence shows substantial crowds despite inclement weather. *Le Progrès de la Côte-d'Or* reported that '[a]ll of Dijon and a multitude of visitors' attended (25 March 1935, p. 3). The *Bien Public* reported 'the first day was a triumphal success' – replete with an 'atmosphere of truculent joy as may be found in Burgundy where we know how to celebrate traditions by paying homage to

Figure 3. Mère-Folle hanging after her mock trial during Dijon's 1935 carnival. Photograph by permission of *Pays de Bourgogne.*

Burgundian products, especially its marvelous wine' (*Bien Public*, 25 March 1935: 1). A participant, Gérard Pérès, wrote to Dijon's City Hall to say that he had been happy to see 'the crowds regain a childlike happiness in these civilized times' and that the 1935 Mère-Folle had 'proved that it is possible to animate and stir people out of their lethargy' (letter from Gérard Pérès dated 11 April 1935 to Gaston Gérard, Archives Municipales de Dijon, 1I1/164, 'Fêtes 1935').

Aside from Suzanne Perri's assault of Louis Démoulins with an iron bar, whose connection to the violence mimed by carnival puppets we can only speculate about, Dijon's 1935 Mère-Folle did not threaten prevailing social order. The Police Commissioner reported only 'lost and found' items in his log for 27 March. These included: one key purse, a silk scarf, Claude-Eugène Boucher's wallet containing his Veteran Card (*carte de combatant*), a small wallet with Edmond Rougeot's calling card, a small bonnet, City of Dijon dog tags, two men's bicycles, a bank note, a small child's bracelet, a woman's woollen shawl, a domesticated rabbit, a string of keys, a *cache-nez*, and a pair of gloves ('Épaves et trouvés', 1935, p. 3). On 25 March, when a radiator cap thrown from a third-floor window injured a pedestrian's foot and a small child's jaw was broken by a kicking horse, all that remained at the end of the day were: one man's bicycle, a child's fur, one *cache-nez*, several wallets, a man's hat, a woman's fur, two birth certificates, an ear ring, a small package of laundry, and a musical instrument. The cultural work accomplished by the Mère-Folle lay in a series of gender reversals calculated to buttress established gender norms.

A six-day calendar of activities for the '1935 Mère-Folle Festival' may be reconstructed from the Committee of Festivities' preliminary notes. It includes the selection of royal officers, various forms of masculine competitions, processional parades, musical entertainments, prankish and exuberant behavior, an abundance of food and drink, as well as the symbolic righting of a world turned upside-down reminiscent of Medieval and Renaissance carnivals. Father Good Times (Père Bontemps) invited the crowds to participate in Dijon's celebrations. Invoking the notion of order and a happy polity, Père Bontemps implored – in pseudo-Renaissance-speak – the city's 'good, happy, and pleasant natures' to attend while enjoining the 'hypocrites, old farts in slippers, sorry grotesques, and pimply-faced urchins to take their meanness elsewhere' ('Père Bontemps…', 1935, p. 3). These *jouissances* (pleasures) were held in two public squares and inside Crazy Mother's Dwelling (Logis de la Mère-Folle) temporarily erected in front of city hall ('Preliminary Program', Archives Municipales de Dijon, 1I1/ 164 'Fêtes 1935 et 1936').

The Logis de la Mère-Folle was a 500 square meter structure similar to pavilions assembled for Dijon's annual Gastronomical Fair. Recalling the 'massive eating' that accompanied early modern festivals, the Logis accommodated spectacles and balls where 'entertainment and dancing' remained 'animated through the night', as well as, a general banquet promoted as serving 'all the specialties of Dijon and the best Burgundian vintages' for 1000 paying guests. Dijon's *Bien Public* covered the carnival events and reported on the evening's entertainments. Performers at the banquet included the comedian Jilune, the dancers Gina and Ginette, the baritone Fred Gouin, the singer 'Fabienne', and the acrobat Jim with his balancing dog (*Jim et sa chienne, equilibriste*). Indeed, much of the Mère-Folle festival's success stemmed from its ability to reconfigure cultural signs otherwise perceived as incompatible.

Following the week of festivity and excess, carnival's intended victim, Crazy Mother (Mère-Folle), was publicly tried, condemned and burned in downtown Dijon on 28 March 1935. Bales of hay were lit beneath her body as it hung from a scaffold built for the occasion. 'Joyous flames', a reported called them, shot up the sides of her blue and purple velour dress and 'devoured her darkening body' ('La Mère-Folle allait mourir', 1935b, p. 5; see Figure 4). Amidst clouds of smoke, Mère-Folle swung back and forth until the ropes gave and she collapsed to the ground. '[M]usic played, a chorus was raised, and joyous *travestis* danced *farandoles'*. There followed an 'immense confetti battle' through the streets of Dijon for which 'the crowd's joy knew no limits' (*Le Progrès de la Côte-d'Or*, 29 March 1935, p. 1). The 'return' of Dijon's ribald Mère-Folle festival shows how a traditional practice could still effect 'complex perceptual, cognitive, emotional, ideological, or behavioral consequences' in twentieth-century France. Carnival provided Burgundians an opportunity – through direct participation – to recode and enact a loose script in order to pantomime preferred notions of hierarchy and parody contested practices.

Carnival (the Dijonnais used the term carnival in 1935 and carnaval thereafter) is a festive season of ludic excess occurring immediately before and in preparation for the Christian Lent. Traditional European carnivals choreographed elaborate charivari-like processions in which dramaturgy and masking marked the overturning of daily life. Carnival typically concluded with a final farce that was equivalent to a ritual of public defamation via mock serenade. Much like early-modern renditions of the world-turned-upside-down-and-restored that were staged through improvised parades of oversized, stock characters, Dijon's revived Mère-Folle pageantry was composed of central larger-than-life puppets and costumed characters. A 'gracious' Queen of Chalon, selected from Dijon's sister city, wore a silver dress with a train of white swan feathers and attended by two 'ladies-in-waiting' wearing pink and

Figure 4. 'So Dijon, we still know how to have fun in our city. Let everyone be ready to participate in the gaiety of the last night of the feast' (*Bien Public*, 27 March 1935, p. 3). Photo showing Mère-Folle's burning effigy during Dijon's 1935 carnival. Photograph by permission of *Pays de Bourgogne.*

carrying flower bouquets in Mère-Folle's traditional colors of yellow and red. The Queen's passage was marked by the crowds' 'warm acclamations'. She served to represent ideal standards of ideal femininity. Following in the procession, Mère-Folle was easily recognized by her 'opulent form' calculated to underscore festive rule and passion and the lower faculties. Her 'gigantic effigy' was drawn by a team of horses that wore Dijon's municipal seal. These sharply contrasting models of ideal femininity and grotesque deformity and were followed by an accompanying troupe of indecorous and disturbingly ambiguous '*travestis* dancing and singing with a devilish air' (*Bien Public*, 25 March 1935, p. 3). Mère-Folle's husband, Father Good Times, followed carrying a large stick with which to 'inflict serious punishment to Lady Crisis [Mère-Folle]' and thereby correct misrule, reclaim his wife, and restore order ('Quand ressuscite la Mère-Folle à sa bonne ville Dijon', *Le Progrès de la Côte-d'Or* 25 March 1935, p. 2). Mère-Folle was followed by a storm of confetti containing her court, entourage, local dignitaries, a band of fools, animated musicians, and assortment of floats that included: a Rabelaisian homage to Burgundian cuisine; Terrot, the 'Big Nursemaid' pushing an enormous baby carriage; a marching jazz orchestra; 'Bacchus', the miner's 'Secret Train'; and merchants' floats promoting regional products (such as Primat's Cassis and Houdart's mustards), archetypical carnival effigies irreverently mocked authority, pantomimed allegories of misrule righted, and engaged in mock battles (*Bien Public*, 22 March 1935, p. 3) through which to sublimate any remaining need for freedom from habitual constraint or lingering hostility.

Katherine Frank's discussion about how individuals (re)negotiate and contest their socio-cultural identities by enacting alternative interpretations of ideologically prescribed social roles is pertinent to carnival costume and mask practices. Frank identifies several categories that allow us to consider the spectator's relation to the inhabited spectacle. She suggests that individuals develop an awareness of identity through performative categories equivalent to 'being' the role by 'suspending disbelief' in the traditional myths; 'doing' the role by recognizing the 'façade' and nonetheless assuming control of its performance along preferred ideological lines; 'resisting' through feigned ignorance or ridicule; and, most importantly, 'subverting' the paradigm by 'twisting, imitating, and parodying traditional scripts ... in a very public, polished display' (Frank, 2002, p. 178).

Vital exuberance, the potential threat to social order, and the ultimate containment of excessive consumption, sexuality, and violence are themes central to traditional carnival culture (Burke, 1978, pp. 186–187) and relevant to inter-war French concerns. In traditional carnival, a fat man and thin woman were cast to represent the cultural opposition between bawdy carnival and chaste Lent:

> Lent was a time of fasting and abstinence – of abstinence not only from meat but from eggs, sex, play-going and other recreations. Hence it was natural to represent Lent as emaciated ... as a kill-joy, associated with the cold-blooded creatures of Lenten diet. Whatever was lacking in Lent was naturally emphasized in Carnival, so that 'Carnival' was represented as young, cheerful, fat, sexy, a mighty eater and drinker, a Gargantuan or Falstaffian figure. (Burke, 1978, p. 188)

This binary opposition was projected into everyday relations between men and women. 'The *topos* of the woman-on-top was one of the most enjoyed' during carnival (Davis, 1975, pp. 128–129). This reversal or world-turned-upside-down topography represented women's insubordination and insolence as responsible for chaos, disorder, excess, and, consequently, the violence necessary to restore order. The image of the unruly woman offered a powerful, multivalent counter-weight to patriarchal order. The manipulation of traditional images of the unruly woman, argues Natalie Zemon Davis, 'is partly a chance for temporary release from traditional and stable society; but it is also part of the conflict over efforts to change the basic distribution of power within society' (1975, p. 131). The exaggerated dramaturgy of carnival lent itself to a pedagogy concerning national virility and appropriate gender roles in 1930s France.

Contemporary regional literature confirms that traditional devices were used to contain gender's 'intractable demands'. As the evolution of early twentieth century rural occupations entailed significant labor realignments, none perhaps, was perceived as more threatening to patriarchal notions of national virility than the growing presence of women in the labor force (Segalen, 1984). One Burgundian novel in particular, Gaston Roupnel's *Nono,* elevated the village farce into an enduring allegory about the struggles of modern couples in an attempt to diffuse anxieties concerning female liberation (Roupnel, 1910, pp. 136–137, 2006, pp. 173–198). Roupnel employed *charivari,* public defamation, and mock serenade, to police adulterous sexuality and define woman's responsibility for domestic order, the household economy, and rural labor (Roupnel, 1933, pp. 126–127; Childers, 2003; 2001; Muel-Dreyfus, 2001). Roupnel's cautionary tales buttressed the conservative politics of an insecure French Republic that blamed depopulation on feminism (birth control and abortion) and countered the threat of women's suffrage with paternalist 'family allowances' by playing to its collective socio-erotic fears. Modern Burgundians raised carnival in its most traditional form in order to police contemporary women through Mère-Folle's mock trial and execution.

Indeed, Mêre-Folle's bawdy behaviour recalls Laura Kipnis' notion of the improper body as a mode of social resistance (Kipnis, 1999; Featherstone, 1991, p. 86). Her proto-feminist disobedience, wanton sexuality, and uncouth rowdiness widened behavioural options and threatened political resistance. Sublimated through festive ritual, the momentary usurpation by the lower orders and organs during carnival week (consider the entrance to Mère-Folle's Logis in Figure 5, p. 486) was ultimately rejected through the burning (as in Mère-Folle's effigy) of the unruly woman on the festival's final day. Consistent with contemporary pressures on 'New Women' to acquiesce before familial and state demands, carnival's rejection of gender equality represents an inability of successive regimes throughout the 1930s to either support progressive politics or reject France's culturally conservative heritage (Faure, 1989; Noiriel, 1999; Dalisson, 2007; Whalen, 2007b).

The fortification of threatened gender roles is suggested by a further reversal. Dijon's Mère-Folle represents carnival and Lent simultaneously. In contrasting the 'thin old woman, dressed in black and hung about with fish' against raucous carnival as 'a fat man, pot-bellied, ruddy, cheerful, often hung about with eatables ... seated on a barrel, or accompanied by a cauldron of macaroni' (as seen, for example, in Brueghel's 1559 paintings of the *Battle of Carnival and Lent*; see, Burke, 1978, p. 185), Dijon's Mère-Folle's grotesque size, excessive behavior, and evident body functions (consider the entrance to the Mère-Folle's Logis) (see Figure 5) compounds her role as an undisciplined woman deserving to be burned at the stake like an early-modern witch (in contrast to the male effigies burned during carnival in eighteenth-century Paris (Garrioch, 1986, p. 197) and not directly related to the burning of Santa Claus in effigy – in front of 250 children – by the Catholic Church in Dijon on 24 December 1951 (Lévi-Strauss, 1952). The semiotic conflation of desirable and odious qualities of carnival and Lent in the form of Mère-Folle compounds this reversal. The resulting mother/whore hybrid grotesquely suggests a second inversion in which ideas associated with nourishment are displaced downward, toward Mère-Folle's lower organs. Hence, the spect-actors who entered Mère-Folle's Logis through the entrance located near her sexual organs individually enacted a collective rape of the village prostitute as metonyme for unruly and intractable women.

Mère-Folle's aesthetic and dramaturgy staged a kind of state-controlled public *jouissance.* Different agendas (gastronomy, labour, gender, etc.) were in this way layered to redefined public spaces wherein 'several modes of ritual invention operate side by side and produce different types of relation between festival and social reality'

Figure 5. 'Grotesques like masks embody cultural contradictions'. Slavoj Zyzek, *On the Sublime of Ideology* (1989), p. 28. Photograph showing the entrance to Crazy Mother's Dwelling (Logis de la Mère-Folle) from Archives Municipales de la Ville de Dijon, 'Mère-Folle' (series 6Fi 269).

(Martin, 2001, p. 5) that Stallybrass and White would call 'a kind of educative spectacle' (1986, p. 38). Opulently saturated with excessive, over-determined, and exaggerated markers, carnival's heteroglossic logic identified tensions while uniting Dijon's individual subjects/citizens within a shared civic experience.

Carnaval as carnival's rebirth

> A cultural mutation without doubt, and mutation toward the cultural that marked its originality and unity of the period [Popular Front], as brief as it was. (Ory, 1994, p. 22)

> Opening the doors of culture to the people did not guarantee that they would select the culture being offered. (Jackson, 1988, p. 131)

Whereas the foregoing has shown how traditional carnival practices recreated under modern circumstances can curb existing and/or emerging cultural practices, the following aims to show how they could also be used to transform socio-political disenchantment into civic solidarity and regional identity.

The Popular Front acquired a mystique of political unity, social optimism, spontaneous play, and cultural transformation likened to the 'experience of life without hierarchy' (Davis, 1975, p. 103). Dudley Andrew's and Steven Ungar's interest in whether or not those attending the 'political drama' of the Popular Front experienced it in a 'straightforward' way (Andrew & Ungar, 2005, p. 11) leads me to question the extent to which Dijon's carnival articulated, investigated, negotiated, and/or left unresolved contemporary ideological tensions, social contradictions, spatial contestations, utopian agendas and personal evasions. Describing the theatrical politics – processions, strikes, marches, simulated funerals, rallies, mock trials, floats with elaborate *tableaux vivants* – employed under the Popular Front, Julian Jackson opines that 'theatrical imagery came naturally to observers and participants' who described their experiences as 'playing a role' and 'living in a Pirandellian dream world' (Jackson, 1988, p. 103). Indeed, the Popular Front's 'most glorious moments', writes Julian Jackson, 'were lived in the street' (p. 114). What activated audiences was 'less the content of the play than the fervor of the audience, less the nature of the art than the relationship between audience and the spectacle' (p. 127).

As with Dijon's previous administration under Gaston Gérard, Dijon's new Socialist mayor, Robert Jardillier, endeavored to diffuse political differences by supporting an eclectic array of social and cultural agendas. Indeed, Julian Jackson had argued that the Popular Front 'was concerned less with creating new cultural forms, with the *content* or artistic creation, than with the democratization of existing, traditional culture which it hesitated to call into question' (Jackson, 1988, pp. 120, 126–127). One can therefore hazard the presumption that carnival markers were neither fixed nor determined. In fact, the themes and elements found in post-1935 carnaval parades blend interdependently and do not therefore point exclusively to modern, urban and industrial or pre-modern, folkloric, and traditional contexts and practices. Although, for instance, Dijon's Communist Party had singularly opposed the 1935 carnival on the grounds that it reflected bourgeois and commercial rather than proletarian interests, it reversed its original position in 1937 when it sought to revive the 'old Dijonnais tradition of Mère-Folle' to ridicule regional targets and political personalities in 1938 (cited in Poirrier, 1995, pp. 729 and 738; see also Poirrier, 1991, p. 63).

By managing the civic representations of the cultural contradictions inherent in French modernization, Dijon's cultural intermediaries – much like the butchers and fishmongers who scripted their economic interests into early-modern carnival and Lenten battles (Bristol, 1989, pp. 75–81) – cleverly influenced how they would be received (Laferté, 2006; Whalen, 2008). The 1932 *Guide des Fêtes Populaires et Traditionelles de la Bourgogne* (Guide to Burgundy's Popular and Traditional Festivals), for example, reveals a willingness to blend traditional and modern cultural traditions in a reassuring and modern fashion. Gaston Gérard, for example, appears to personally invite Burgundians and tourists to avail themselves of the 'colorful manifestations' of the regions popular and traditional festivals (1932, p. 1). Celtic fires, summer theater, costume revivals, wine festivals, vernal, equinox and solstice festivals, Saints' days, gastronomic fairs, country *ébaudes*, musical recitals, dance troupes, and literary awards all competed for the crowds attending Burgundy's annual festivals. This practice of giving familiar folk idioms new resonance is evident in Max Cappe's 'Mère-Folle's Song' (scored by Jean Dagey and performed on 23 May at Dijon's 1935 carnival). His lyrics privileged motifs and themes associated with the community of Burgundian vintners ('Beaux Masques', 1935, p. 3). As with his weekly contributions to *La Vie dijonnaise*, Cappe's work was popular precisely because it embraced both traditional and contemporary cultural agendas.

The Popular Front closely associated itself with innovative artists. Among these was Jacques Copeau, the director of the Vieux Colombiers Theater in Paris (Jackson, 1988, p. 129; Morin & Richard, 2008; Leroy & Roche, 1986). Copeau's productions

Figure 6. Carnival float celebrating wine-making traditions during Dijon's 1935 carnival. Photograph by F. Perrodin. Permission by the Musée de la vie bourguignonne Perrin de Puycousin, Dijon.

addressed traditional themes through rustic productions (with titles like 'The Wash-house', 'The Snails of Burgundy', and the 'Ballet of the White Wine and the Red Wine') staged at wine festivals throughout Burgundy and France during the late 1920s and early 1930s ('Jacques Copeau en Bourgogne', 1928, pp. 253–255; Bouaké, 1926, p. 1). A reviewer of the Burgundian Minstrels' (Les Ménestrel bourguignons) choral repertoire noted that '[p]opular', rather than traditional, songs 'unite men, eliminate differences, and confound origins' (Heuvrard,1934, p. 4490). Singing 'songs of the provinces, old songs of former days', was promoted as a way of 'penetrating the popular soul' (Jackson, 1988, p. 128). Dijon's Socialist mayor and esteemed musicologist, Robert Jardillier, for example, supported popular, traditional and classical choir groups and wrote about the links between their musical traditions (Jardillier, 1926, pp. 567–581; Emmanuel, 1923, pp. 349–371). In March 1937, Jardillier led the University of Dijon's choir through a repertoire of traditional songs that explored 'the forgotten origins of popular songs from diverse French provinces' (Thomas, 1937, p. 6). He also created the Cercle Rameau to revive interest in the music of Jean-Philippe Rameau, a late Baroque composer born in Dijon and best known for his harpsichord, opera, and theatrical works (1683–1764) and reviewed diverse – including highbrow – literature favourably while in office as a Socialist representative (Jardillier, 1927, p. 4). These simultaneous cultural agendas reveal Dijon's musical interests as ranging more broadly than ideological preoccupations with rural/urban and traditional/popular antinomies would otherwise suggest (Ory, 1994, p. 503, 508) (see Figure 6).

The Popular Front is also remembered for having pursued a regimen of unity, strength, health, and happiness (*bonheur*) through recreational activities and the democratization of sport culture (Rearick, 1997, pp. 209–211; Pénet, 2006; Berstein

& Milza, 1991, pp. 86–87; Callède, 2007). Sporting spectacles appealed to rural, village, and urban populations equally. The French organized themselves into all manner of sporting societies, clubs, corporations, fraternities, unions, syndicates, or chambers according to their political, social, professional, cultural, industrial, agricultural, peer, gender, age, or religious affiliations. Both Dijon's Cyclodrome and Parc des Sports, for instance, were built under the center-right mayorality of Gaston Gérard during the early 1930s. The foot-races, water jousting, tugs-of-war, wrestling matches, and egg races of early-modern festivals and carnivals (Burke, 1978, p. 184) were echoed in Dijon's 1935 'Mère-Folle Challenge', which pitted the French and Belgian boxing and cycling teams (*Le Progrès de la Côte d'Or*, 24 March 1935, pp. 1, 5). The Auxonnais Cycling Club entered its 'Fêtes Artistique et Carnavalesque du Club Cycliste Auxonnais' float in Dijon's carnival from 1936 until 1939 (*Le Progrès de la Côte-d'Or*, 4 March 1936 and 7 March 1939).

Carnival provided the possibility of redirecting contemporary anxieties into a collective, polymorphous mirth that few could monopolize (Andrew & Ungar, 2005, p. 174). The basic topics and themes of the 1936 carnaval remained consistent between 1937 and 1939. The carnaval parade ('Grand corso carnavalesque') of 1936 entailed a greater number and diversity of floats ('up to thirty') that followed a route that included working-class neighborhoods beyond Dijon's historic and commercial downtown area, most parade floats remained surprisingly eclectic as ideological markers (see Figure 7). Floats were subsidized and built by assorted neighborhoods, professional clubs, commercial organizations, industries, merchants and a variety of unions. Dijon's Terrot cycle manufacturer again contributed oversized puppets riding assorted wheeled contraptions. The town of Chenôve provided a float of traditionally

Figure 7. Carnival float celebrating industrial labor during Dijon's 1935 carnival. Photograph by F. Perrodin. Permission by the Musée de la vie bourguignonne Perrin de Puycousin, Dijon.

attired ('de vendangeurs et de layottes d'allure bien bouguignonnes') grape harvesters; the Goujon neighborhood echoed traditional Lenten imagery with an enormous sardine float surrounded by fishermen in modern attire and equipment; the Maladière neighborhood reproduced the architecture and monuments of the Place La Fontaine, while another float, 'accompanied by the sound of hurdy-gurdy, violin and cello' depicted a 'Morvan Wedding' composed of a couple traditionally attired. The Automobile Club of Burgundy entered a smoking old jalopy driven by two tourist/chauffeurs wearing thirty-year-old attire. A restaurant championed the city's gastronomic reputation by floating an enticingly set table attended by duly appreciative guests. The Electricity Company's three 'House of Electricity' floats – in the form of an enormous radiator, home iron and domestic boiler each attended by workers – represented 'all the advantages this modern fairy brings to our interiors'. The Gaz de France's oversized lamppost waddled along accompanied by uniformed lamplighters, meter readers, and bill collectors. The Company added a second float depicting the advantages gas provided in the kitchen, represented by an enormous woman cooking at a modern stove. The 'Family and Work' float entered by the Committee for South Dijon did not aspire to utopian and scientific socialism. It rather conventionally promoted a 'hen-that-lays-the-golden-egg' theme for its working class denizens: a metal-smith working his forge figured signified a 'good life assured through industriousness'. The Worker's Sport Union's float featuring campers 'cooking or resting as they would on an isolated plateau' around a well equipped tent. French chauvinism was reflected in the Poussots community's 'Normandie de Suzon' float, a scaled replica of the French Line's record-breaking ocean-liner Normandie. Commandeered by a gold braided captain and escorted by seamen hurling 'serpentine' braids into the crowds, the 'Normandie de Suzon' indecorously towed old pots, rent shoes, torn umbrellas, 'and other objects encountered along its journeys' off its stern. 'Small Commerce is ruined, crushed, and strangled' by 'excessive fees, the economic crisis and high taxes' read the placards on another particularly penurious looking entry. The 'Death of the 48-Hour Week' offered the most popular and eye-catching entry as laborers sawed off the nose (representing eight hours) of an oversized crocodile. More surprising, but not inconsistent with the use of black-face and *tirailleurs sénégalais* at other events, was the float, 'accompanied by general laughter', representing 'Négus Hailé Sélassié' outrageously seated in a baby carriage and whipping a black slave who pulled him along. This was accompanied by a less racially coded treatment in the France-America Good-Will float with Dijon's mayor, Robert Jardillier, wearing a mask of Franklin Roosevelt. Accompanied by the usual carnival assortment of picturesque *travesties* dancing, gesticulating, and singing joyfully, the Grand Corso followed a horseman outfitted in *Ancien Régime* finery ('Le grand Corso carnavalesque de Dijon', *Le Progrès de la Côte-d'Or*, 9 March 1936, p. 4).

Popular Front additions to the carnaval parade in 1937 included a beehive shaped Forty-Hour Week, the Trumpet Corps of Dijon, dancing Marianne on the Republican float, the factory shaped Rebirth of Work float 'symbolizing the joyous activity of a smoking factory'. 1938's new floats addressed traditional themes: a Youth Hostel's (Auberge de la Jeunesses du Cercle Laïque) oversized rustic hut surrounded by Pierrot puppets, the Char du Morvan depicting an expanded traditional Moran family accompanied by accordion players, the Building Trades Association's Char du Surarbitre' pulled by three oversized Burgundian snails (*Le*

Progrès de la Côte-d'Or, 9 March 1936, p. 4, 8 March 1937, 14 March 1938; *La Bourgogne Républicaine*, 14 March 1938, pp. 1–2, 4–5, 20 March 1939, p. 4).

The consistency of motifs shared by Dijon's carnival and subsequent carnavals shows that both events supported a variety of political and cultural agendas. Indeed, addressing those who might wish to think otherwise, Dijon's First Vice Mayor assured readers of the *Bourgogne Républicaine* in 1937 that the prevalence of traditional elements supported the event's 'democratic nature' (Thomas, 1937, p. 6). By the time of the 1939 carnaval, the Duchess of Burgundy's float was pulled by a bottle of wine and a pot of mustard (*La Bourgogne Républicaine*, 20 March 1939, p. 5, 21 March 1939, p. 5).

Conclusion: the politics of Carni/aval

One year after Mère-Folle's execution, the politically left-leaning and previously supportive *Progrès de la Côte-d'Or* reproduced pictures of parade floats from Paris' 1936 Lenten (Mi-Carême) celebrations but said nothing about Dijon's 1935 revival. A brief column discussed the 'Origins of Carnival' and solemn processions associated Dijon's sixteenth-century Society of Mère-Folle while remaining silent about the riotous behavior of early-modern carnival celebrations which explicitly 'attack[ed] both legitimate authority and illicit sexual activity' (*Le Progrès de la Côte-d'Or,* 21 March 1936, p. 5; 'Les origins du Carnaval', *Le Progrès de la Côte-d'Or*, 19 March 1936, p. 4; Bristol, 1989, p. 69). Had the revival of carnival been too provocative for Dijon's bourgeois neighborhoods already annoyed by the licentiousness of the annual Student's Ball? (Gérard, 1960, p. 175). Did the severe economic downturn prompt the celebration's demise? Sales of Dijon's 1936 carnaval 'Program', however, were sufficient to send 'numerous needy children on green vacations in the open air' ('Triomphe du Carnaval Dijonnais', *La Bourgogne républicaine*, March 1937, p. 6). Unfortunately, no extant deliberations or pronouncements from Dijon's Committee on Fairs and Festivals or Chamber of Commerce address this point. We simply know that Dijon's new, pro-Popular Front municipal administration scuttled carnival in 1936 and replaced it with a less traditionally defined and less centralized carnaval in the cities of Dijon, Chalon and Mâcon.

Political divisions intensified in France in 1936 and 1937 as a conservative Senate resisted the Popular Front's progressive legislation, economic conditions worsened, military preparedness taxed existing resources, and French international alliances left many confused about Leon Blum's commitments. The ensuing stalemate at the national level translated into increased and factional competition for cultural leadership at the regional and local levels (Palau, 2005, pp. 225–233; Lévêque, 1994, p. 209). Dijon elected the Socialist candidate, Robert Jardillier as mayor in the spring of 1936 while Chalon retained the socialist-communist front already established in May 1935. Not surprisingly, neither city, now under Popular Front mayors, decided to repeat the success of the 1935 carnival just before the 1936 elections. Was this because the Popular Front feared the threat of poorly controlled streets during a period of intense political rivalries or that it did not have a cultural policy sufficiently different from its predecessor?

Mère-Folle's final execution, I would argue, underscored the success of an ambitious, broad-based cultural agenda that anticipated and, for that reason, limited the Popular Front's ability to formulate an alternative cultural agenda for Dijon's citizenry in 1936. Philippe Poirrier has identified the 1935 Mère-Folle carnival as 'the last witness of a 'cultural politics' narrowly linked to a political economy supported locally by important industrial and commercial interests' (Poirrier, 1995, pp. 377, 389). Reflecting on the conflictual management-labor relations of late 1930s Dijon, Poirrier also investigates whether or not there existed a cultural divide between pre- and Popular Front interests can be detected in Dijon's carnival and carnaval parades (10 M 81-89 'Grèves', Archives départementales). Examining the same, I find that the thematic and topical continuities between carnival and carnaval, from 1935 through 1939, trumped coherent ideological agendas which no political party could afford to claim while retaining the diversity of votes necessary to form coalition governments. This is consistent with Julian Jackson's re-evaluation of the Popular Front's cultural policies in 1936 as surrounding 'not the restrictive "proletarian" but the all-embracing "*populaire*"' in order to define the nation's 'cultural patrimony as generously as possible'. Rather than resulting from properties inherent in either festival format, the demise of carnival and the longevity of carnaval resulted from changing political regimes that, while first appropriating and settling on carnaval, were able to infuse either with similar content. Spectacles such as Dijon's 1935 Mère-Folle carnival – despite its political provenance in the conservative administration of Gaston Gérard – were readily absorbed by the new regime whose own Minister of Education, Jean Zay, sought to promote educational policies that provided a 'broad general education [and] not in any sense a proletarian culture' (Jackson, 1988, pp. 122–123).

If carnival can provide 'an opportunity for assessing how citizens construe their relationship with power' (Martin, 2001, p. 18), what then was the political function of carnival and carnaval in 1930s Dijon? Both events provided local authorities with an effective, time-tested method of 'reuniting the individual with the collective' to strengthen feelings of solidarity (Bristol, 1989, p. 29; Rearick, 1985, p. 131). Both events highlight the persistence and continuity of popular practices (such as masking, inverting, indifference, irreverence, etc.) while underscoring how they intensify, reconfigure and otherwise employ the carnivalesque at any historical moment. Both mobilized existing passions ('joyeux turbulences populaires' and 'réjouissances générales'), choreographed support for civic agendas ('les tonifiantes joies des amusements sincères et honnêtes'), distracted attention from failed or unpopular policies, and prompted individuals to identify with the 'collective enthusiasm and good spirits' (Thomas, 1937, p. 6).

Rather than adopting either a stable definition of carnival or chasing endless polysemic possibilities in order to determine carnival's utility across Dijon's changing political climate between 1935 and 1937, it may be wisest to recognize that carnivals structure a range of pre-discursive affinities, semiotic associations, and performative behaviors whose political significance lay in their ability to conjure and provisionally mobilize familiar cultural ties and desired symbolic associations. Seen in this light, it is my contention that the Burgundian inter-war regional cultural movement may be said to have momentarily mediated the seemingly irreconcilable demands of modern and traditional practices. It packaged and promoted a regional identity that held out the promise of preserving traditional values alongside aggressive modern marketing strategies. This reconfiguration of Burgundian collective identity required the collaboration of the region's cultural intermediaries, private investment, and political leaders to direct the region's productive forces and patterns of consumption through fairs and festivals designed to model and enact those practices. In an era no less politically charged than the early-modern France, diverse social actors in the streets of Dijon during the late 1930s could seize, in Natalie Zemon Davis' words, 'upon older social forms and change them to fit their needs' (Davis, 1975, p. 122). The success of such festivals survived the 1930s to inform French cultural politics through the remainder of the twentieth century.

References

Aching, G. (2002). *Masking and power: Carnival and popular culture in the Caribbean.* Minneapolis: University of Minnesota Press.
Agulhon, M. (2003). Le Centre et la périphérie. In P. Nora (Ed.), *Les Lieux de mémoire* (pp. 2889–2906). Vol. 2. Paris: Gallimard.
Archives départementales de la Côte-d'Or, series M and SM.
Archives Municipales de Dijon, 1I1/164, Fêtes.
Baker, C. (2002). *Making sense of cultural studies.* Thousand Oaks: Sage Publications.
Bakhtin, M. (1984a). *Problems of Dostoevsky's poetics.* Manchester: Manchester University Press.
Bakhtin, M. (1984b). *Rabelais and his world.* Bloomington: Indiana University Press.
Bazin, J.-F. (1997). *La Bourgogne.* Rennes: Editions Ouest-France.
Beaux Masques. (1935, March 26). *Le Progrès de la Côte-d'Or,* 3.
Bercé, Y.-M. (1994). *Fête et révolte, des mentalities populaires du XVIe au XVIIIe.* Paris: Hachette.
Berstein, S., & Milza, P. (1991). *Histoire de la France au Xxe, 1930–1945.* Bruxelles: Editions Complexe.
Bien Public, Le. (1935–1939).
Bouaké, M. (1926, August 30). *Bien public* (p. 1).
Bourgogne d'Or, La. (1925–1933).
Bourgogne républicaine, La. (1937–1939).
Bristol, M.D. (1989). *Carnival and theater.* New York, Routledge.
Burke, P. (1978). *Popular culture in early modern Europe.* New York: Harper & Row.
Callède, J.-P. (2007). Maitres et ministres entreprennants l'invention d'une politique des sports (1919–1939). In P. Tétart (Ed.), *Histoire du sport en France du Second Empire au régime de Vichy* (pp. 175–181). Paris: Vuibert.
Cappe, M. (1932). *Les Chants du terroir, poèmes bourguignons.* Dijon: Lèpagnez.
Cappe, M. (1938). La confrérie des Chevaliers du Tastevin. *La Bourgogne d'Or,* September–October: 226–231.
Carcano, M. (1997). *Dijon à travers les ages.* Dijon: Damidot Frères.
Childers, K. (2003). *Fathers, families, and the state in France, 1914–1945.* Ithaca, NY: Cornell University Press.
Dalisson, R. (2007). *Les Fêtes du Maréchal.* Paris: Tallandier.
Davis, N.Z. (1973). The rites of violence: Religious riot in sixteenth-century France. *Past and Present, 59,* 51–91.
Davis, N.Z. (1975). *Society and culture in early modern France.* Stanford: Stanford University Press.
Debord, G. (1995). *The society of the spectacle.* New York: Zone Book. (Original work published 1967)
Demossier, M. (1999). *Hommes et vins: Une anthropologie du vignoble bourguignon.* Dijon: Presses Universitaire de Dijon.
Demossier, M. (2006). *Nono,* ou la construction du vigneron comme archétype de la culture locale. In A. Bleton-Ruget & P. Poirrier (Eds.) *Le Temps des sciences humaines* (pp. 173–198). Paris: Manuscrit-Université.
Dudley, A., & Ungar, S. (2005). *Popular Front Paris and the poetics of culture.* Cambridge: Belknap Press.

Dyer, R. (1981). Entertainment and utopia. In R. Altman (Ed.), *Genre, The musical: A Reader* (pp. 175–189). London: Routledge.
Emmanuel, M. (1923). La Chanson populaire: Emprunt au folklore national. *La Revue de Bourgogne, 13*, 349–371.
Épaves et trouvés. (1935, March 28). *Le Progrès de la Côte-d'Or*, 3.
Farge, A. (1993). *Fragile lives: Violence, power and solidarity in eighteenth-century Paris.* Cambridge, MA: Harvard University Press.
Faure, C. (1989). *Le Projet culturel de Vichy: Folklore et révolution nationale, 1940–1944.* Lyon: Presses Universitaires de Lyon.
Featherstone, M. (1991). *Consumer culture and postmodernism.* London: Sage Publications.
Fiske, J. (1989). *Understanding popular culture.* Boston: Unwin Hyman.
Frank, K. (2002). Stripping, starving and the politics of pleasure. In L. Johnson (Ed.), *Jane sexes it up* (pp. 171–207). New York: Four Walls Eight Windows.
Garrioch, D. (1986). *Neighborhood and community in Paris, 1740–1790.* Cambridge: Cambridge University Press.
Gérard, G. (1932). *Guide des Fêtes Populaires et Traditionelles de la Bourgogne*, pp. 1–2.
Gérard, G. (1935, October 31). Les fastes et liesses de Dijon. *Bien public*, pp. 1 and 4.
Gérard, G. (1959). *Le Miroir du coin et du temps.* Dijon: États Généraux de la Gastronomie Française.
Gérard, G. (1960). *Dijon, ma bonne ville... souvenirs et confidences.* Dijon: États Généraux de la Gastronomie Française.
Golan, R. (1995). *Modernity and nostalgia: Art and politics in France between the wars.* New Haven: Yale University Press.
Guide des fêtes et populaires et traditionalles de la Bourgogne. (1930–1934).
Guilard, C. (1999). 'La Foire Gastronomique de Dijon'. Université de Bourgogne: Diplôme D'études Approfondies.
Heuvrard, A. (1934, July). Les Ménestrel bourguignons. *Le Mirroir dijonnais, 15.170*, 4490–94.
Holt, M. (2000). Election festivals, power and society in sixteenth-century Burgundy. In K. Friedrich (Ed.), *Festive culture in Germany and Europe* (pp. 21–29). Lewiston, New York: The Edwin Mellon Press.
Horkheimer, M., & Adorno, T. (1995). *The dialectic of enlightenment.* New York: Continuum. (Original work published 1972)
Horne, A. (2004). *The seven ages of Paris.* New York: Vintage Books.
Inglis, F. (1988). *Political culture and political power.* New York: St. Martin's Press.
Jackson, J. (1988). *The Popular Front in France: Defending democracy, 1934–38.* Cambridge: Cambridge University Press.
Jacques Copeau en Bourgogne. (1928). *La Bourgogne d'Or*, 253–255.
Jacobs, J. (1996). *Edge of empire: Postcolonialism and the city.* New York: Routledge.
Jameson, F. (1981). *The political unconscious.* Ithaca, NY: Cornell University Press.
Jardillier, R. (1926). Pour une musique bourguignonne. *La Revue de Bourgogne, 16*, 567–581.
Jardillier, R. (1927, July 30). Deux Livres par Gaston Roupnel: *Siloë – Hé! Vivant!. Le Progrès de la Côte-d'Or*, 4.
Kellner, D. (2003). *Media spectacle.* New York: Routledge.
Kipnis, L. (1999). *Bound and gagged.* Durham: Duke University Press.
L'inauguration de la XIième Foire Gastronomique. (1931, November 1). *Le Progrès de la Côte-d'Or*, 2.
La Mère-Folle allait mourir. (1935a, March 29). *Le Progrès de la Côte-d'Or*, 1.
La Mère-Folle allait mourir. (1935b, March 29). *Bien Public*, 5.
Laferté, G. (2002). La Production d'identité territoriales à usage commercial dans l'Entre-Guerre en Bourgogne. *Cahiers d'économie et sociology rurales, 62*, 66–95.
Laferté, G. (2005). Un 'folklore' pour journalists: la Confrérie des Cavaliers du Tastevin. *Ethnologie comparées, 8*, 1–32.
Laferté, G. (2006). *La Bourgogne et ses vins.* Paris: Belin.
Le Retour de la Mère-Folle. (1935, March 23). *Le Progrès de la Côte-d'Or*, 1
Le Roy Ladurie, E. (1979). *Carnival: A people's uprising in Romans.* New York: G. Braziller.
Leroy, G., & Roche, A. (1986). *Les Écrivains et le Front Populaire.* Paris: Presses de la Fondation nationale des sciences politique.

Lévêque, P. (1994). *Histoire de Chalon-sur-Saône.* Dijon: Editions Universitaire de Dijon.

Lévi-Strauss, C. (1952). Le Père Noël supplicié. *Les Temps modernes, 77*, 1572–1590.

Martin, D.-C. (2001). Politics behind the mask: Studying contemporary carnivals in political perspective, theoretical and methodological suggestions. *Questions de recherché/Research in question, 2*, 2–34.

Mayne, J. (1994). Paradoxes of spectatorship. In L. Williams (Ed.), *Viewing positions: Ways of seeing film* (pp. 155–183). New Brunswick: Rutgers University Press.

Moentmann, E.M. (2003). The search for French identity in the regions: National versus local visions of France in the 1930s. *French History, 17*(3), 307–327.

Morhain, G. (2009). Les années Gaston Gérard. *Bourgogne magazine*, January, 34–36.

Morin, G., & Richard, G. (2008). *Les deux France du Front populaire: Chocs et contre-chocs.* Paris: L'Harmattan.

Moriss, P. (1935). La Mère-Folle de l'Infanterie dijonnaise. *La Bourgogne d'Or*, May, 179–189.

Muchembled, R. (1985). *Popular and elite culture in France, 1400–1750*. Baton Rouge, LA: University of Louisiana Press.

Muel-Dreyfus, F. (2001). *Vichy and the eternal feminine.* Durham, NC: Duke University Press.

Nicolaÿ, F. (1902). *Histoires des croyances, superstitions, moeurs et coutumes.* Paris: Victor Retaux.

Nicole, H. (1933, August 16). Les Fêtes balladoires. *Progrès de la Côte-d'Or*, 1–2.

Noiriel, G. (1999). *Les Origines républicaine de Vichy.* Paris: Hachette.

Ory, P. (1994). *La Belle illusion: Culture et politique sous le signe du Front Populaire, 1935–1938*. Paris: Plon.

Palau, P. (2005). *Histoire du département de la Côte-d'Or.* Dijon: Darrantière. (Original work published 1978)

Patouillet, X. (1971). *L'Emeute de Lanturelus à Dijon en 1630*. Université de Dijon: Mémoire de Diplôme.

Payne, J.M. (1998). The politicization of culture in applied folklore. *Journal of applied folklore, 35*(3), 251–277.

Peer, S. (1998). *France on display: Peasants, provincials, and folklore in the 1937 Paris World's Fair.* Albany, NY: State University of New York Press.

Pénet, M. (2006). *Été 36 sur la route des vacances.* Paris: Omnibus.

Père Bontemps ... (1935, March 23). *Le Progrès de la Côte-d'Or*, 3.

Poirrier, P. (1991). 'Les Heures bourguignonnes:' Une relecture communiste de l'histoire locale. 1937–1939. *Gothiques, 27*, 63.

Poirrier, P. (1994). Le Retour de la Mère Folle et les fêtes carnavalesques à Dijon (1935–1939). In A. Corbin, N. Gérôme & D. Tartakowsky (Eds.) *Les Usages politiques des fêtes aux xixe et xxe siècles* (pp. 377–389). Paris: Publications de la Sorbonne.

Poirrier, P. (1995). *Municipalité et culture au xxe siècle: Des beaux-arts à la politique culturelle.* Université de Bourgogne: Thèse.

Progrès de la Côte-d'Or, Le. (1935–1939).

Quand ressuscite la Mère-Folle à sa bonne ville Dijon. (1935, March 25). *Le Progrès de la Côte-d'Or*, 1–2.

Rearick, C. (1985). *Pleasures of the Belle Epoque: Entertainment & festivity in turn-of-the-century France.* New Haven, CT: Yale University Press.

Rearick, C. (1997). *The French in love and war.* New Haven, CT: Yale University Press.

Reid, D. (2001). Carnival in Rouen: A history of the Abbaye des Conards. *Sixteenth Century Journal, 32*(4), 1027–1055.

Roupnel, G. (1910). *Nono.* Paris: Plon-Nourrit.

Roupnel, G. (1922). *La ville et la campagne au xviie siècle: Etudes sur les populations du pays dijonnais.* Paris: Editions Ernest Leroux.

Roupnel, G. (1933). Comment les Françaises ont fait la France. *La Nouvelle revue féminine, 3*, 126–131.

Roupnel, G. (n.d.). Dijon. In *Fonds Roupnel* (n.p.). Maison de Sciences de l'Homme, Université de Dijon.

Schwartz, V. (1998). *Spectacular realities: Early mass culture in fin-de-siècle France.* Berkeley, University of California Press.

Scott, J. (1990). *Domination and the arts of resistance.* New Haven: Yale University Press.

Segalen, M. (1984). *Marie et femme dans la société paysanne.* Paris: Flammarion.
Shaya, G. (2004). The Flâneur, the badaud, and the marketing of a mass public in France, circa 1860–1910. *American Historical Review, 109*, 41–77.
Shuldiner, D. (1998). 'The politics of discourse: An applied folklore perspective'. *Journal of Folklore Research, 35*(3), 189–201.
Stallybrass, P., & White, A. (1986). *The politics and poetics of transgression.* Ithaca, NY: Cornell University Press.
Thomas, L. (1937, March 8). La Mi-Carême de Dijon. *La Bourgogne républicaine,* (pp. 1 and 3).
Urry, J. (1995). *Consuming places.* London: Unwin.
Vachon, P., & Gerley, J. (1992). La Mère-Folle de Dijon: Fête de la satire et des satyres. *Pays de Bourgogne, 155*, 1–9.
Weber, F. (2000). Le Folklore, l'histoire et l'état en France 1937–1945. *Revue de synthèse, 4*(3), 453–446.
Whalen, P. (2007a). 'A merciless source of happy memories': Gaston Roupnel and the folklore of Burgundian *terroir. Journal of Folklore Research, 44*(1), 21–40.
Whalen, P. (2007b). Republican consumer culture and Burgundian cultural regionalism at the 1937 Paris Exposition. *Cultural Analysis, 6*, 31–62.
Whalen, P. (2008). 'Revenir aux saines traditions de la vielle France': Gastronomie et tourisme bourguignon à l'époque de Gaston Gérard. *Annales de Bourgogne, 79*(3), 259–280.
Williams, L. (Ed.). (1994). *Viewing positions.* New Brunswick: Rutgers University Press.

Ethnic identity and elite idyll: a comparison of carnival in Buenos Aires, Argentina and Montevideo, Uruguay, 1900–1920

Kristen McCleary

History Department, James Madison University, Harrisonburg, USA

This article compares the transformation of carnival celebrations in Buenos Aires, Argentina and Montevideo, Uruguay at the beginning of the twentieth-century. It argues that changes in carnival practices in the River Plate region, a geography shared by these two national capitals, cannot be understood without taking a comparative stance since the demise of carnival in Buenos Aires was linked to the rise of a vacation culture in Montevideo. In addition, the more dramatic urbanization process experienced by Buenos Aires and the longer political stability of the nineteenth century here enabled elite politicians, commercial interests, and the middle and upper classes to regulate carnival behaviors to such an extent that the festivities became increasingly lackluster. In contrast, the slower urbanization process in Montevideo combined with the government's close but not overzealous regulation of carnival, allowed Montevideo's carnival to flourish. The disparate nature of Montevideo's carnival celebration that included the involvement of the city's diverse neighborhoods was also an important aspect of this carnival's popularity and success. This article also assesses the historiography on Montevideo's carnival which has cast a negative light on the impact of modernization and the festival. In comparison to Buenos Aires', carnival in Montevideo was fomented by governmental regulation. Finally, this article examines the relationship between carnival and each of the city's African-descent populations.

> Montevideo sustains the ancient custom of keeping carnival, masked and with illuminations, flower-throwing and costumed *corsos* (parade routes), in a fashion which entirely throws into the shade, the now moribund carnival in Buenos Aires… Thousands of *porteños* cross the river each carnival season to enjoy Montevideo's elaborate and unrestrained street celebrations. (Ross, 1916)

Carnival in the Americas has most often been studied as a legacy of the Afro-Latin American cultural heritage. The most vibrant carnival celebrations have and continue to occur largely in those regions or nations with significant African-descent populations, such as the Caribbean (Bettelheim & Nunley, 1988; Aching, 2002; Scher, 2003) and Brazil (Da Matta, 1991; Ergood, 1991; Crook, 1993; Chasteen, 1996; Dunn, 1999). This article explores the ways in which carnival experiences in two countries known more for their inhabitants of European descent, rather than African or indigenous populations, became transformed in the early part of the twentieth century. It shows how carnival in two South American capital cities transformed as a result of the process of modernization and urbanization that took place most markedly at the beginning of the twentieth century. Ultimately, this article seeks to understand why and how carnival celebrations faded away in Buenos Aires, Argentina, while they continued to flourish on the other side of the River Plate area, in Montevideo, Uruguay, a transition that took place most notably between 1910 and 1915.

Carnival in these capital cities, which ranged from three days to an entire month of pre-Lenten celebrations, experienced similar histories as it moved from largely popular unrestrained street celebrations during the colonial eras to increasingly commercialized and regulated ones between 1880 and 1920. Despite sharing similar historical trajectories, however, the carnival celebrations of Montevideo, Uruguay have shown great stamina and resilience, rebounding from moments of political repression, notably during the dictatorship of 1973 to 1985 (Remedi, 2004; Xosé de Enríquez y Vidal, 2004). In contrast, the carnival celebrations of Buenos Aires have experienced the opposite trajectory, declining almost to the point of disappearance through much of the twentieth century. The municipal government of Buenos Aires only recently re-introduced two days of vacations for city employees during the days of carnival in an effort to stimulate tourism linked to carnival's global popularity. In addition, the province of Buenos Aires recently voted in an ordinance that for many years had been the cornerstone of the city's anti-carnival policy – legalizing the throwing of water and paper on the city's streets during carnival (*The Economist*, November 2005). While *The Economist* blamed Buenos Aires' moribund carnival on the military dictatorship of the 1970s, this article argues that one needs to look back further to understand how carnival faded away here. Ultimately, this article argues that the disparate transformations of carnival in the Río de la Plata region can be traced back to the early 1900s, when Argentine elites began to forego their own celebrations in favor of those across the Río de la Plata, a hundred and thirty miles away, in Montevideo. The demise of carnival in the early 1900s in Buenos Aires coincided with and contributed to the expansion of such celebrations in Montevideo.

Most studies of carnival celebrations in their many guises have tended to focus on national histories and the ways in which these annual pre-Lenten festivities either defied or reified the existing social order (Bakhtin, 1984; Stallybrass & White, 1986). In this volume, Piers Armstrong summarizes the complex theoretical history and debates about the role of carnival and the carnivalesque, emphasizing the polyvalence of carnival in the Americas:

> Carnival presents contrary energies. On the one hand, it is a centripetal rite whose conservative effect in reiterating and implicitly celebrating existing order is greater than the contrary centrifugal effect of opening up space for dissidence. On the other hand, in the long term the conservative effect is strongly countered – not only by conspicuous and conscious dissidence, but also by the gradual assimilation to the national mainstream of imported references and local 'bottom-up' expressions.

Similarly, carnival in the Río de la Plata had fluid meanings which shifted depending on the historical era, demographic reality and dominant power structure. The contours of carnival were often controlled by a regulatory elite political body which, as Armstrong suggests, co-opted elements of the lower classes. Yet, the diverse expressions of carnival also permitted resistance and response to the elite's version of the ideal celebration. In Buenos Aires and Montevideo, the meaning of carnival changed over time but it always embodied a negotiated series of rituals that were shaped and contested by political and economic elites, the burgeoning middle classes, commercial interests, and the popular classes. A recent spate of scholarship has focused renewed attention on carnival in Argentina and Uruguay as a way to gain insight into the modernization of their capital cities and the concomitant transformation of social relationships in the era following national independence (Alfaro, 1991; Anderson, 2005; Andrews, 2007; Chamosa, 2003a, 2003b; Chasteen, 2004; Remedi, 2004; Seigel, 2000). These scholars show how the meanings of carnival in the Río de la Plata region were continually debated, redefined, and contested amongst social classes, ethnic groups, and urban regulators during the nineteenth century.

Part of the way in which carnival presents a complicated ritual with shifting meanings is the fact that carnival, itself, embodies a series of events, performances, and rituals. Each component of the celebration may contrast in meaning with other parts. In the Río de la Plata, carnival activities ranged from unregulated street play where citizens engaged in pranks of throwing water, or confetti-filled or rotten eggs on unsuspecting passers-by, to official parade routes designed, implemented and regulated by urban elites, to masked carnival balls in the commercial spaces of theaters, to carnival groups parading, unregulated, through suburban neighborhoods. In most of the Río de la Plata region, this pre-Lenten festivity in both cities drew a great deal from Carnevale in Venice, Italy with continual reference to the stock Commedia dell'Arte figures of Colombine, Pierrot and Harlequin. In Montevideo, in particular, the celebrations were personified around the figure of Momo, the carnival king, and the Marqués de las Cabriolas (Marquise of Capers) who ceremonially opened carnival, in Montevideo. Momo was ritually 'buried' during a final celebration. As will be discussed below, these European antecedents melded and merged with Afro-Latin American music to form a new hybridized cultural expression arising from the complex power relationships and inter-actions between the different ethnic groups that shape Latin America (Chasteen, 2004; Andrews, 2007).

While carnival experiences have largely been studied as national histories, in the case of the Río de la Plata, one cannot fully understand how carnival practices became transformed over time without taking into account the larger regional geographical and demographic context. Here, carnival celebrations in the early twentieth century were shaped by the transforming relationship between Montevideo and Buenos Aires. Transnational, more than national, concerns shaped carnival into a point of destination for *porteños* (residents of the city of Buenos Aires) to visit Montevideo, as Uruguay and its southern beaches transformed into a tourist destination.

Justification for comparative value

Uruguay and Argentina and their capital cities have followed similar historical trajectories and, indeed, during the colonial era belonged to the same administrative region, known as the Río de la Plata. During the colonial era, neither city was prominent. The first notable division between the two regions occurred in 1776, when

Buenos Aires was chosen to be the seat of the viceroy of the Río de la Plata, an area that included the territory of Argentina, Bolivia, Paraguay and Uruguay. This administrative decision placed Buenos Aires in a dominant position, and since then it has outnumbered its eastern counterpart in population and economic activity. In the national era, both Argentina and Uruguay followed similar paths, achieving their independence from Spain around the same time. Montevideo and Buenos Aires were established as their respective national capitals due to their ports, which stimulated economic and demographic growth. As port cities, they both received a number of European immigrants at the end of the nineteenth century, and saw their already relatively small African-descent populations decrease in size and visibility as a result of the flood of primarily Spanish and Italian arrivals (Andrews, 1980, 2004). Both cities shared the same vision at the turn of the century: that the progress of the nation would be symbolized through the material development of its national capital (Needell, 1995; Rosenthal, 1996).

Despite the similarities between these cities, Buenos Aires' population always greatly outnumbered that of Montevideo, a point crucial to contrasting the disparate trajectory of these cities' carnival celebrations. At the turn of the century, the population of Buenos Aires averaged about four times that of Montevideo. In 1884, for example, Montevideo's population numbered 164,028 (Rilla, 2005) and in 1887 Buenos Aires had about 433,373 inhabitants (Moya, 1998). Both cities grew dramatically over the next thirty years, due to the arrival of thousands of immigrants to their shores, primarily from Spain and Italy. In 1908, the year of Uruguay's national census, Montevideo had a population of 309,231, and Buenos Aires had a population of 1,176,480, about four times greater.

In addition to demographic differences, these southern cone nations also had divergent experiences in terms of late-nineteenth century political stability. Francisco Panizza details Uruguay's political disruptions during this era, noting that 27 men occupied the presidency in the hundred or so years between independence and José Batlle y Orodoñez's second presidency from 1911 to 1915 (Panizza, 1994). In addition, the country suffered civil wars in 1897 and 1904. While Argentina experienced some political instability at the same time, Panizza argues that, in general, almost 70 years of constitutional continuity prevailed. In terms of carnival, Argentina's political stability enabled city regulators to exert control over the local shape of carnival (Panizza, 1997). In Montevideo, lack of political stability resulted in a lack of political interference in carnival that allowed it to retain a shape defined and embraced by its most numerous practitioners, the working classes (Andrews, 2007). When political stability prevailed, during the era of Batlle, Montevideo's carnival benefited from an organizational structure that supported its growth and did not try to constrain it. In relationship to carnival, Gustavo Remedi shows that Batlle began to foment public performances by 'using community money to build [neighborhood] stages', beginning in 1904 (Remedi, 2004, p. 64).

Carnival regulation in Buenos Aires

As Patricia Anderson convincingly argues, carnival 'reflected the tensions that underlay [Buenos Aires'] growth and transformation. Instead of drawing attention to the sophistication of the city and its inhabitants, the festivities highlighted the deep divisions that threatened the attainment of progress and "civilization"' (Anderson, 2005, p. 40). Indeed, the demise of carnival in Buenos Aires can be seen as resulting from these tensions, exacerbated even more by the intense urbanization demands and conflicts of the late nineteenth century. During the post-colonial era, carnival evolved from a popular street activity to one that was increasingly regulated and restrained, symbolized by its transformation from predominantly outdoor to largely indoor celebrations. The spatial transformation of carnival had deliberate social class ramifications as it was removed from the street, a public space accessible to all social classes, into commercial theater buildings, areas that economically included primarily the middle and upper-middle social classes. Almost all of the political elite of Buenos Aires shared a disdain for the outdoor festivities, resulting in prolonged attempts to restrain, regulate, and control them. By 1920, the previously well-known and popular street activities of carnival here, where *comparsas* (carnival groups) joined *corsos* (parade routes) on the city's streets, had faded away.

Buenos Aires' elite had attempted to control pre-Lenten 'excesses' dating back to the colonial era. Commonly, the earliest administrative concerns about carnival were clearly linked to the celebratory activities of the city's African-descent populations. Carnival celebrations provided a visible and lasting venue from which African Nations, the name given to membership organizations created by African ethnic groups in the Americas, blended their own ritual festivals and cultural expressions with those of the culturally Christian peoples who first enslaved them in Latin America. That is, Catholic practices, such as the celebration of religious festivals and rituals, were one of the ways in which Spanish and Portuguese slave-holders attempted to acculturate their African-descent populations to the dominant mores of their belief system. These same festivals provided a social space in which local African traditions could also be incorporated, resulting in a synchretic mixture of African and Christian rituals. George Reid Andrews and John Charles Chasteen have both noted the deep African roots of carnival celebrations throughout the Americas. (Andrews, 2004, p. 123).

During the colonial era, slaves in Buenos Aires routinely held a variety of social activities that were often related to religious activities resulting from their being organized into *cofradías*, or lay brotherhoods, by the church. Black dances were often conflated with religious holidays, especially carnival (Andrews, 1980, p. 161). There were several times when colonial administrators issued regulations attempting to prevent African slaves from paying homage to Momo, the lord of carnival, by prohibiting traits of their celebration such as dancing and drumming (Andrews, 1980, pp. 156–157; Puccia, 1974, pp. 57–68).

Despite an attempt to curtail African-descent peoples and traditions in carnival festivities, the role of African cultural practices continued to be an important component of carnival in Argentina until the latter part of Juan Manuel de Rosas' rule. A regional strongman whose close ties with the popular classes and strong arm techniques enabled him to 'restore' order to the nation, Rosas was well-known for supporting the festivities of the popular classes (Lynch, 1981). Despite this reputation, Rosas himself prohibited carnival celebrations during the final years of his rule, 1845 to 1852. Ironically, he employed the same language that those before him and those after him would use in order to stifle carnival celebrations – these unrestrained festivals were violent; they damaged public order and the good customs of the people; and they threatened to destroy an 'enlightened' society. Rosas also felt these celebrations were contrary to a hard-working people, that they damaged the economy, and were unhygienic (Puccia, 2000, p. 94). Patricia Anderson argues that once Rosas had consolidated his political rule he no longer needed carnival celebrations to reach out to the popular classes. John Charles Chasteen notes this as a moment of important transition since it was after the Rosas era that carnival in Buenos Aires reflected the interests of the elite rulers over the popular classes (Chasteen, 2000, p. 49).

Carnival in the era of modernization

By the onset of the era of modernization in 1880, which coincided with Buenos Aires becoming the nation's capital, the governing and popular attitudes towards carnival were long established: The city's inhabitants continued to embrace the polyvalent forms that carnival took, while city administrators lamented the lack of control displayed by the city's revelers. Beginning in the 1880s, complaints about the African cultural traditions of carnival celebrations had been replaced by a new fear: that carnival could unleash urban disorder in this epoch of rapid demographic growth. Between 1880 and 1920, carnival in Buenos Aires would no longer belong to the popular classes but would be absorbed and experienced most fully by the middle and upper social classes. Carnival became an important way in which these social groups made manifest and consolidated their social status.

The strong ideological stance of Argentina's elite and its emphasis on reforming the national capital explain why this era focused so intently on taming carnival celebrations. Elite attempts to shape Buenos Aires into the Paris of South America drew upon the ideological proclivity of statesmen like Domingo F. Sarmiento who opposed 'civilization' with 'barbarism', claiming that civilization occurred in the cities and all that was barbaric took place in the countryside (Sarmiento, 2003). As a result, especially after 1880, when Buenos Aires was chosen as the site of the national capital, the city served as an important symbol of the nation's 'progress'. City administrators wished to remake Buenos Aires into the Paris of South America by destroying remnants of the city's colonial past and introducing new wide avenues based on Georges Haussmann's designs for Paris. Jeffrey D. Needell has noted that this attempt to remodel the city was also meant to 'represent (or induce) the change in public consciousness' of its citizens (Needell, 1995, p. 519). Given the intellectual project of this elite whose attention centered on Buenos Aires as a showcase of material progress, it was no wonder that carnival celebrations were a recurring problem, a reminder of the city's 'unfortunate' past as an uncivilized backwater.

While the era of modernization saw carnival celebrations becoming more regulated, especially as the city council began to exert greater control over parade routes, Buenos Aires' carnival retained its most problematic feature, that of the *entrudo*, the practice of throwing water, and other objects such as flour, confetti, and streamers with small stones embedded in them, onto pedestrians. Such pranks resulted in violent confrontations between 'prankster' and victim (Puccia, 2000, pp. 72–73). The *entrudo* was not native to Argentina. It was also the subject of great consternation in the nineteenth-century in Argentina's neighboring countries, Brazil and Uruguay. For example, Sandra Lauderdale Graham gives a vivid description of the *entrudo* (meaning, she notes, 'to mock or make fun of') in early nineteenth-century Río de Janeiro (Graham, 1992, pp. 66–71). She remarks that these pranks were ways in which the lower classes could claim and make the street, or public space, their own by subjecting all social classes to their surprise attacks. In Argentina, traveler accounts invariably complain about the annoyance and outright danger of unrestrained carnival celebrations during this era:

> Carnival is the event of the year. Everything looks forward to it. Everything stops for it. For weeks preceding [carnival], the windows of business houses were filled with masks of all descriptions and other appliances for its proper observance. Chief among these appliances is the *pomo*, a small can or bottle of soft, flexible tin, with screwed cap over the small neck. The *pomo* contains perfumed water. Every variety of perfume is discernible. It is said that poisoned *pomos* [tubes filled with stagnant water] are used as a means of taking vengeance on an enemy or settling an old grudge. The top being removed, and the *pomo* squeezed through the finger and thumb, a fine, steady stream of water is poured upon the object of attack. The chief aim is at the eye... The many murders committed in retaliation caused the interference of the civil power. (Clemens, 1886, p. 174)

The majority of discussions and debates around carnival practices centered on these water games, or *juegos de agua*, as the *entrudo* was called in Buenos Aires. Over the years, seemingly countless pieces of regulation were debated, promulgated, overturned and re-introduced once again in largely unsuccessful attempts to control mayhem on the city streets which inevitably followed the aftermath of the throwing of water onto passersby. For example, water vials were first prohibited in 1876. In 1881, they were legalized, although it was still illegal to throw water onto passersby from anything larger than an eight-inch vial, a law that targeted the frequent use of jars and buckets. Carnival pranks also intersected with public health concerns especially in 1887 when a cholera epidemic hit the city. City officials outlawed even the water vials which were imported from England, and many charged that the vials tubes contained unhygienic dank water. In addition, the city regulated that property owners would be held responsible for any objects thrown from their premises (Buenos Aires, 1887, pp. 472–474).

The upper-class broadsheet, *La Gaceta Músical* (20 February 1887) described the impact of this legislation as follows:

> Water, merry-makers, and *pomos* have been shut down this year by our tyrannically great police; no one will show their heads during the day, nor throw water, or do any of the other thousand 'tacky', entertaining and characteristic things our carnival is known for. Decorative lights, painted figures, flowers and the immense human wave of the population surrounded by the streets Victoria, Peru, Florida, Rivadavia and Piedad [streets of the main downtown area] will not turn Buenos Aires into a universal Babel during three days of carnival fun. The Judas Errante is not permitted. Only the clubs will celebrate carnival. *Momo* instead of being dressed in cardboard sandals and a flecked mess of colored clothes, will wear a velvet mask, don suits of satin, or sport a tuxedo.

The quote gives insight into the transformation of carnival from a popular street celebration to an event more designed for the elite, as it becomes privatized into a celebration for social centers.

As carnival grew increasingly commercialized, a group of manufacturers wrote a petition to the city council urging them to re-think their ban on these vials (Buenos Aires, 1901, November 15). In their petition to the city council, they explained that the current prohibition regarding the sale of these items was anachronistic, since it was only in the past when the police were 'indifferent to carnival activities' that the vials caused problems. In fact, they creatively argued, the legalization of vials would actually improve public safety, since liquids squirted from a vial were not as likely to cause injury as those thrown down from above via pails and buckets.

Debates over the reintroduction of water vials, a seemingly frivolous topic, shed light on deeper concerns about carnival on the part of its regulators. Indeed, the city council's views of carnival intersected with three primary issues: hygiene, public order, and commerce. Sometimes these concerns overlapped. Debates regarding disorder easily turned to philosophical discussions about the role that carnival should play in the city and in the nation. Those members who argued for the legalization of the sale of water vials and for the retention of carnival argued that historically people had sought to amuse themselves in a variety of ways that had not always met the approval of their rulers. Using the positivistic language of the time that stressed the importance of Europe as a center of civilization, one council member argued that most 'civilized' parts of the world had 'barbaric' traditions: 'Even France, a beacon of civilization, has bullfighting in Nime and Nantes. Furthermore, do we really want to imitate the tyrant Rosas who for a decree in the year 1845 prohibited carnival in the entire nation in all of its forms?' (Buenos Aires, 1900, January 22).

The council members also decided to prohibit serpentines, streamers manufactured in France and England, that were rolled up and thrown into crowds. Many of them were thrown with a hard object, like a stone, wrapped in the middle. The streamers were also visible reminders of the disorder of carnival since they littered the city's streets for months, and they reportedly interfered with the functioning of phone lines, since so many were thrown onto wires. While many council members had felt that outlawing streamers was a bit excessive, since every 'civilized' country in the world allowed these streamers to be thrown, they were nonetheless prohibited. Councilmember Varela, for example (jokingly, one assumes) declared that he would vote against the prohibition of serpentines, 'because I do not want to be more civilized than the Europeans' (Buenos Aires, 1900, January 22).

In addition to restricting the use of water vials and confetti, the city council also increasingly regulated the location and length of parade routes within the city. As the city grew in population, the central streets were increasingly congested. In 1905, for example, the city passed an ordinance which prohibited carnival parade routes from passing through the city's center. The magazine, *La Ilustración Sud-Americana* (20 February 1905), commented that the current year's carnival celebrations would 'not be much' due to increased regulations. Increased traffic in the era of the automobile contributed to congestion exacerbated by carnival celebrations. Many newspapers reported that this ordinance was the one that ultimately killed carnival in Buenos Aires. *La Razón* (10 January 1910, Buenos Aires), for example, noted that Momo 'died in Buenos Aires not only a victim of his own vices but also due to an inopportune municipal ordinance'.

Of course, city regulations would not mean much if there had been no attempt to enforce them. The period between 1880 and 1910 shows a marked increase in police surveillance and presence during carnival. The police's primary role was to maintain order along the various carnival parade routes. As early as 1881, police had detailed directions about how to monitor parade routes in all twenty police districts of the city (Buenos Aires, 1881). Internal police publications note the lack of success in this arena since they had to threaten individual agents with the loss of their jobs if they were seen dancing while on duty, suggesting, of course, that such occurrences were not uncommon (Buenos Aires, 1923, p. 57). This image of a lackadaisical police force transformed into one that had a reputation for overzealously carrying out its role. Between 1900 and 1915, it became increasingly common to read complaints directed towards the police in relation to their carnival duties (7 February 1910, *La Razón*, Buenos Aires, p. 3). Municipal funds helped increase the number of police on duty during carnival, and individual agents even received overtime pay. Many press accounts questioned why the Buenos Aires police force could not achieve what police departments everywhere else had supposedly managed to do – maintain public order without choking all of the life out of festivities (1 and 7 February 1901, *La Razón*, Buenos Aires, p. 3).

Overall, increased city regulations and their strict enforcement transformed carnival in Buenos Aires by ending street festivities and moving carnival into the interior social spaces of theaters and social clubs (12 February 1910, *La Prensa*, p. 2). Attending theater was already a well-established custom in the city, which had over twenty theaters operating at any one time. Profit-driven theaters were well positioned to receive carnival revelers as well, since they regularly hosted 'special event' evenings throughout the year, which garnered larger entrance fees. Carnival, that most special of celebratory events, offered lucrative opportunities. Indeed, enterprising managers quickly extended carnival balls on weekends to cover the entire month. Because carnival coincided with the summer months of the southern hemisphere, the off-season for theater performances, theater managers made the most of this opportunity to increase their box office sales by hosting masquerade balls.

Beginning in the late 1890s, the transition of carnival celebrations into indoor theaters was frequently noted in the press. For example, in 1893, *La Ilustración Sud-Americana* (18 February, p. 3) remarked that theaters were places of great fun and animation during carnival, with the exception of the Opera where 'our golden youth have caused a drunken spectacle'. On 8 February 1910, *La Prensa* reported, 'The scarcity of parade routes has caused people to look for their entertainment in dancing that took place in the city's theaters'. In the early 1900s, the central role of theaters in carnival celebrations had been further cemented due to the rise of dance craze fads in Buenos Aires such as the Cake Walk and tango (McCleary, 2002, pp. 223–243; Puccia, 2000).

Carnival also became increasingly marked as an expression of growing consumerism and the commercialization of leisure time activities. Theaters, rather than parade routes, were social spaces able to integrate the commercial dictates of the time with carnival. For example, carnival began to be defined by contests in which participants would compete for money prizes awarded for best costumes and best dancing. Responding to the pressures of competitions, contests and commercialization, homemade costumes gave way to store bought ones, and people were encouraged to buy all sorts of carnival paraphernalia that they had previously done without or produced at home. Magazines and newspapers began carrying advertisements for all sorts of carnival merchandise in addition to costumes, such as confetti, serpentines, masks and other costume accessories.

The local press contributed to the consumer frenzy by publishing photos of children who showed up at theaters and social clubs to compete for costume prizes.

Indeed, contests seemed deliberately to target children who came dressed up as the era's typical characters such as ballerinas, gauchos, gypsies, dominoes, Pierrots, Colombines, and washerwomen. Upper middle class magazines, such as *Fray Mocho* and *Caras y Caretas*, regularly devoted a few pages to photo spreads of costumed children. The transformation of carnival costumes also reflected the Europeanization and 'Argentine-ization' of these celebrations. The sad clown Pierrot, his object of desire, Colombine, and Harlequin, the object of Columbine's desire, were well-known figures in European carnival balls, and they dominated the adult costumes in Argentina. Children, however, provided a unique space for the society to incorporate costumes of Argentine identity, notably gauchos. Both transformations, however, were clear moves away from the Afro-Argentine inspired costumes and cultures of carnivals past. Photographs of carnival during the pre-1910 era show Afro-Argentines and European-Argentines dressing up in stereotypical 'mammy' costumes quite frequently.[1] But during the first years of the twentieth century those costumes gave way to European ones and distinctively Argentine ones. Micol Seigel has shown how carnival provided a social space that allowed the numerous immigrants new to the city, the opportunity to become Argentine with the rise of *centros criollos* between 1900 and 1910 (Seigel, 2000, p. 59). Composed largely of immigrants, these criollo groups came together to perform 'authentic' Argentine history, acrobatics, and music.

The increased role of children in carnival celebrations went beyond the transformation of immigrants into Argentines. This newly targeted audience also promised theater owners additional sources for profit. They increasingly held balls for children as part of the relatively new addition of using afternoon times for matinee events. Matinees, in general, targeted women and children as a way to increase theater attendance in the city.[2] Theaters, in sum, were venues that offered to contain carnival celebrations in a limited space that was easier to control than festivities taking place on city streets, and they accommodated and defined the commercial nature of carnival that had now become inextricably linked to a middle and upper-middle class consumer culture.

Rise of carnival in Montevideo

This commercialization of carnival coincided with the development of a summer vacation culture, especially pronounced for the upper and upper-middle classes of Buenos Aires. Summer vacations began to link the citizens of the Río de la Plata together. This was not a two-way exchange, however, as *porteños* frequented Montevideo and its surrounding beaches, rather than the other way around. Montevideo, in particular, responded to the arrival of tourists by developing a relatively solid tourist infrastructure that involved developing water-front hotels and restaurants and also coordinated the development of fluvial and urban systems of transportation to carry *porteños* and other tourists from place to place. Transportation especially was coordinated to accommodate the arrival of thousands of visitors during carnival season. Expanded train schedules were regularly reported in the city's newspapers during carnival celebrations (1914, January 30, *La Razón*, Montevideo, p. 1). The port was also expanded and improved in 1915 to accommodate more visitors. In addition, Montevideo began to promote carnival in other cities through the use of posters. For one season, over two hundred such advertisements were posted throughout Buenos Aires, and fifty each in the Argentine cities of Rosario, Santa Fe, and La Plata, in addition to other cities in Uruguay (1912, February 7, *La Razón*, Montevideo).

The increasing commonality of spending January and February in Montevideo and its beaches including Pocitos, Ramírez, Piriapolis, and, eventually, Punta del Este, was reflected in the coverage of Buenos Aires' magazines such as *Caras y Caretas*, *P.B.T.* and *Fray Mocho*. These publications, targeted to the upwardly mobile middle classes and already 'upward' classes, regularly covered elite social events in Montevideo during the summer months.[3] The development of tourism in Uruguay also coincided with the demise of outdoor carnival activities in Buenos Aires. As a result, summer vacations began gradually to introduce *porteños* to an alternative carnival experience. Gordon Ross, the English journalist who resided in Buenos Aires, and whose epigraph opens this article, gives a detailed account comparing the two capital cities, carnival, and the rise of vacations in Montevideo as a destination for Argentines (Ross, 1916, pp. 79–85), remarking that with its beautiful white beaches, Uruguay should have ambitions to become *the* seaside resort for all of South America. As more and more *porteños* began to leave Argentina for their vacations, the Argentine government and businesses attempted to stem the flow of tourist dollars by offering reduced-price vacations at their national vacation spots, such as Mar del Plata (1917, January 5, *La Tribuna Popular*, p. 4). In particular, a tourism war erupted between 1914 and 1915 as the two national capitals each competed for tourist dollars in light of a diminishing economy resulting from the onset of World War One.

A number of factors resulted in the migration of vacationing Argentines to the beaches near Montevideo. Economic growth in the early twentieth century in Argentina resulted in an increase of *porteños* with both the time and money to vacation. Montevideo quickly responded to a demand in tourism by developing restaurants, beach-front hotels, and transportation systems linking Montevideo to its beach communities. Vacations to these beach towns were promoted in Buenos Aires as a result of an increase in public health awareness that stressed the need for people to seek out 'salt baths' and fresh air. The concept of vacation itself was clearly linked to public hygiene. For example, the Buenos Aires city council often noted the poor air quality of its downtown especially during the summer months when wind and dust made the city almost uninhabitable especially in combination with the modernizing efforts of the city to repave central streets. One traveler also saw the social class tensions that had exploded in Buenos Aires in the 1919 Semana Trágica as a reason for Argentines to vacation in Montevideo:

> The traveling Argentino finds here [in Montevideo] a pleasant haven of refuge from the slings and arrows of his Republic's outrageous socialism, a sport where efficient service and civility are not considered to be beneath the dignity of waiters, porters and housemaids. (Koebel, 1920, p. 229)

The increase in summer vacations was a result of modernization and urban growth, and it was socially and politically promoted within the Argentine capital. Leisure time and the concept of vacation became socially prescribed antidotes for the upper classes to escape from the pressures of living and working in an overcrowded and unhygienic city.

Comparing carnival on both sides of the Río de la Plata

From the early 1900s, the Buenos Aires press had commented on the shifting shape of carnival in their own city, usually in a way that suggested the present celebrations were mere shadows of past ones. Utilizing the positivist language of the time, albeit sarcastically, *La Ilustración Sud-Americana* summed up the state of affairs: 'The newspapers and other official documents say we are progressing and probably this, our dying carnival celebration, is the sign of how much progress we have achieved' (1908, March 15, p. 2). One journalist in Buenos Aires published an 'interview' with the symbolic representative of carnival, Momo. The journalist asked him, 'Sr. Momo, we thought you were in Mar del Plata, Montevideo or Nice this season!' Momo: 'No, I am here dressed up like someone who is bored. It truly is amazing. You people are even capable of killing off the immortal Phoenix!' Journalist: 'But surely you have had some fun, Momo?' Momo: 'I threw some serpentines on the phone wires, I lit a torch on a balcony, and I threw some water at people'. Journalist: 'But all that is prohibited sir!' (1920, February 9, *La Razón*, Buenos Aires, p. 1). Of course, both of these examples and numerous others acknowledge that the city's administrators had finally won their battle to tame carnival celebrations by, in effect, over-regulating it. They had, so to speak, thrown the baby out with the bath water. The baby, however, had landed safely in Montevideo. *La Razón* noted, 'Momo has been thrown out of our lands. Now he feels like royalty in Uruguay where he is treated well' (1910, January 31, *La Razón*, Buenos Aires, p. 1).

Montevideo's *La Razón* ran a series of articles in 1914 explaining the reasons why carnival in Buenos Aires was 'dying'. The articles responded to Buenos Aires city council attempting to revive carnival there by overturning the ordinance that had prohibited parade routes from taking place in the city's center (1914, February 14, p. 1). Uruguayans explained that carnival in Buenos Aires had died, in part, because *porteños* had a tendency to import ideas and fashions and, as a result, overlooked and undervalued their own cultural traditions. In addition, the newspaper said that carnival could not be celebrated in such a large city as Buenos Aires, which had a population of 'strangers' and was all 'chaos and confusion'. These traits were anathema to the warm camaraderie that formed a necessary part of authentic carnival celebrations, such as those that occurred in Montevideo. Rising worker unrest furthermore fueled the chaos of carnival on the other side of the Río de la Plata, since the police were not able to control labor strikes and carnival crimes at the same time. In sum, Buenos Aires had itself ruined its own carnival since the city council had tried to eliminate central parade routes, and the population of the city destroyed any semblance of fun by committing acts of violence throughout the days of revelry.

The same newspaper noted a few days later:

> Every city in the world will be celebrating these next three days of carnival. Everywhere that is except for Buenos Aires, a city full of serious men, a boring metropolis, that capital of solemnity that has no room for fun. It is the 19th of February, and you do not see one single preparation for carnival in that great city. You do not even see a single flag. There is no sign that indicates the upcoming days of revelry and exaltation. Carnival is now dead thanks to the declaration of the chief of police that they will be out in full force to supervise the carnival parade route in the central city that the city council has attempted to organize in an effort to infuse life back into the celebration. (1914, February 19)

While *porteños* arrived in great numbers to enjoy Montevideo's beaches and carnival celebrations, it appears that they were most often observers rather than active participants. For example, newspapers often described *murgas* (small groups of carnival revelers) in relationship to the neighborhood from where they originated. There were very few mentions of *murgas* composed of people from Buenos Aires. The one time a *murga* was mentioned as being composed of 'thirty young Argentines', it was noted that the Carnival Commission was aware of that fact, suggesting it was indeed out of the ordinary (1917, February, *La Razón*, Montevideo, p. 2).

Montevideo

Was carnival in Montevideo so different from that of its rival? If so, how and why did carnival differ? Milita Alfaro has shown that carnival in Montevideo underwent transformations parallel to those of the celebrations of Buenos Aires (Alfaro, 1991). The regulations of both cities were surprisingly similar with each city prohibiting water play, the use of religious or military outfits, and carnival revelers from making fun of national institutions, such as singing the national anthem.[4] Alfaro's assessment of carnival in Montevideo during the era of modernization is one in which carnival became restrained, co-opted by the government, and commercialized. She casts these transformations in an almost wholly negative light, drawing from a Foucaultian vision of a society being disciplined and controlled by its regulators.

In comparison to carnival in Buenos Aires, however, one can argue that many of the changes that carnival underwent in Montevideo, here carried out under the administration of José Batlle y Ordoñez' social welfare government, somehow contributed to a more positive carnival experience. The form that carnival took in Montevideo suggests that a negotiation took place in which government, police, and public shaped a carnival that not only was no threat to public order but also offered the public an acceptable form of diversion.

Perhaps the most important feature of carnival in Montevideo was the diversity of activities that took place throughout a wide swathe of the city: *tablados* (neighborhood stages) were built, and people joined carnival societies, attended masked balls at social centers, hotels, and in theaters, went to costume contests, and attended parades during the day and the evening where they could watch allegorical floats, decorated cars and carriages pass by. The activities were always kicked off with a parade that was led by the Marqués de Las Cabriolas (1915, February 15, *La Razón*, Montevideo, p. 2), who was performed by a popular shoe shiner and janitor, named Eduardo Lametz. During his 13-year tenure as King of Carnival, Lametz in his role of Marqués was often a figure of public ridicule. Lametz was a local celebrity, albeit a dubious one, who regularly received press attention. Reporters noted his demands to have protective bars around his float to protect him from all of the objects thrown at him on the annual kick off parade (*La Razón*, Montevideo, 1915, February 15, p. 2). This is significant because Montevideo's festival had curtailed the unruly aspects of *entrudo* not by erasing it altogether but by symbolically transferring it towards the figure of a scapegoat. In this way, street disorder was controlled but the subversive element of social catharsis was retained. Ironically, Lametz' working class status suggests that the middle classes were the group using carnival as a realm of social catharsis.

The press commented on the growth of carnival in Montevideo with about the same frequency they reported on its demise in Buenos Aires. The amount of actual space in each newspaper devoted to carnival coverage greatly expanded between 1910 and 1915. *La Razón* (Montevideo), for example, devoted three full pages of about a ten-page paper to carnival, including the 'above the fold' section on the front page. It also used a reduced-sized font for regular carnival coverage on the days immediately preceding the events, in order to fill in as much detail as possible. One editor commented, 'We confess that we do not have enough space or time to summarize all of the carnival activities that we know of, so we can only offer you the following updates in the space allotted' (1915, February 12, *La Razón*, Montevideo, p. 1). On 23 February 1914, the same paper reported, '*Murgas* are invading the city. Seven boys get together (why are they always groups of seven?) This year there are hundreds of them. From every *conventillo* [tenement] to every street corner'.

The municipality government of Montevideo exerted a great deal of organizational control over carnival in the city, making sure that most of its participants registered with the city's Carnival Commission. Both cities had a number of social centers organizing different carnival activities, but only in Montevideo did the city require all such groups to register their activities with the city government in an effort to provide a larger organizational structure. This organization facilitated the smooth operation of carnival throughout the city's many neighborhoods. Not surprisingly, Buenos Aires' carnival was consistently criticized for its lack of organizational structure.

In addition, the municipal government of Montevideo worked smoothly in conjunction with the city's police to ensure that regulations were streamlined and posted throughout the city. Regulations monitoring carnival were printed annually in the city's newspapers during carnival season as a seemingly friendly reminder. At times the Carnival Commission actually held meetings at the police station, suggesting a fluid relationship between these two entities (1905, February 22, *La Tribuna Popular*, p. 2). Montevideo's police had special training sessions to prepare them for maintaining order during carnival (1905, March 4, *La Tribuna Popular*, p. 3). This implies that Montevideo's city and police understood that the revelry of carnival presented special policing challenges, requiring subtle rather than overt forms of social control.

In direct contrast to the press coverage of policing in Buenos Aires, most of the Montevidean press reported on the actions of the police in relation to carnival in a laudatory manner. For example, after carnival, *La Razón* (1914, February 21, p. 3) reported, 'Police service was impeccable. Parade routes were well organized and the police heads and officers all carried out their jobs very well ensuring public safety. The great number of people and traffic make it difficult, however, to maintain perfect order in the city'. Indeed, the stereotype that carnival in Buenos Aires had deteriorated as a result of the scandalous activities of its own inhabitants seems to have also been embraced by Montevideo's police, who took an active role in apprehending known criminals from Buenos Aires as they disembarked in Montevideo. In 1917, the Montevideo police force was doubled so that agents could identify known *porteño* criminals. The police reportedly detained 16 of these figures and told them they would be under heavy surveillance, if they chose to stay. Most were 'convinced' to return to Buenos Aires.

Role of neighborhoods

One of the most important differences between carnival in Buenos Aires and Montevideo had to do with the disparate nature of Montevideo's celebrations throughout the city's neighborhoods where *tablados*, or neighborhood stages, were an important part of carnival. A commission was first charged with organizing the neighborhood commissions in 1892 (Remedi, 2004). More important, perhaps, was the fact that José Batlle y Ordóñez' government subsidized these stages, beginning in 1904, to gain political support from the popular classes (Remedi, 2004, p. 64). Montevideo's press commented on the funding. In 1914, 50,000 pesos were set aside for the festivities (1914, January 17, *La Razón*, p. 3). This is a notable departure from the practice of Buenos Aires' governmental officials who did very little fiscally to promote *any* cultural practices during the same time period. While the Argentine magazine *Caras y Caretas* noted that working class neighborhood celebrations in La Boca and Flores, for example, were growing in size, these celebrations operated entirely outside of the realm of governmental support. Indeed, it is notable the degree to which the Argentine government (municipal and national) ignored cultural issues outside of the realm of the elite opera house, the Teatro Colón (McCleary, 2002, pp. 362–419).

In Montevideo, *tablados* seemed to arise almost spontaneously in most city neighborhoods, reflecting the great degree to which the *culture* of participating in carnival celebrations had reached there. These neighborhood commissions registered with the city's Carnival Commission and newspapers announced their locations, amount of prizes to be awarded and times of contests. *Tablados* served as more than mere stages in a couple of ways. First, the construction of *tablados* was a serious undertaking, and they were approached as works of art. Neighborhoods competed to win the best *tablado* prizes citywide. Each community organization chose a theme for their *tablado* that was incorporated into the physical design of the stage. Many *tablados*, for example, celebrated the tourism of Montevideo by having a theme such as 'Montevideo's best beaches'. Therefore, the construction of stages was a great undertaking which required that local inhabitants meet over a variety of decisions and then spend a great deal of time together building the stages. The preponderance of *tablados* forced carnival participants to move throughout the city. Rather than gravitate towards the main parade routes in the city center, people who made up carnival groups, moved from *tablado* to *tablado*, from neighborhood to neighborhood, that is, in order to compete for prize monies given out by each neighborhood commission (1917, February 27, *La Razón*, Montevideo, p. 4).

Tablados targeted the two main groups of carnival societies in Montevideo, the traditional *comparsas*, large groups of between 60 and 80 dancers and musicians who dressed up and performed during carnival, and the more informal, *murgas*, who began to appear in carnival celebrations here in the early 1900s. *Murgas* grew in popularity over time while *comparsas* began to decrease in quantity. Originally formed of small groups of seven to eight young men, they easily moved from one *tablado* to another to compete for prizes. They usually had some sort of comedic and oftentimes nonsensical name such as, 'Bad Boys of the Street', 'I want to be Where I was Last Night', and 'Nocturnal Bums'. The names imply that they derive from local, inside jokes, once again reflecting their neighborhood base. Around 1910 newspapers began to devote more and more space to the *murgas* and less and less to the *comparsas*. The smaller size of the *murgas* made it much easier to move throughout the city.

Reading the history of Montevideo through popular press accounts and in comparison to the mordant Buenos Aires celebration offers a new perspective on carnival in that city. Paulo de Carvalho-Neto, Milita Alfaro, and Abril Trigo all present a picture of carnival that stands in stark contrast to the celebration presented in this article. Carvalho-Neto views Montevideo's festival as one which was 'tightly controlled by the government', including censorship that was implemented through the awarding of prizes that reinforced prevailing notions of cultural norms (Neto, 1967). He notes that *comparsas* had to adapt to rules which banned, 'sad or melodramatic scenes and disguises that do not belong to carnival' and 'all indecent words, gestures, or allusions'. Yet in my research, I have found numerous newspaper accounts that disparagingly noted when a particular *murga* or *comparsa* sang especially vivid or controversial lyrics. Reviewing 10 years of newspaper accounts of carnival reported by two of the main newspapers revealed no complaints that the city was heavily exercising its censorship duties. Indeed, the rapid growth of *tablados* and *murgas* between 1910 and 1920 suggests that such an undertaking would have been virtually impossible. The flexibility of *murgas* to move throughout the city and the ability of all of carnival participants to improvise suggests that the city would have a very difficult time exercising its authority in this manner. In fact, if the city had over-regulated carnival it would most likely have suffered a similar fate to the celebration on the other side of the Río de la Plata.

Role of race/ethnicity

Current scholarship on carnival celebrations has stressed the important role that African-descent populations played in maintaining and preserving a vibrant carnival celebration. Did carnival die in Buenos Aires as a result of the loss of a visible Afro-Argentine cultural heritage at the end of the nineteenth century? Did it grow in Montevideo due to the larger presence of Afro-Uruguayans, who remained an important component of carnival? What role, that is, did ethnic and cultural traditions play in preserving the legacy of carnival in the Southern cone? It is problematic to make direct connections between late nineteenth and early twentieth-century carnivals to contemporary ones, since the meanings of these celebrations are inscribed within their particular temporal and social contexts. In fact, many contemporary carnival celebrations in the Caribbean and Brazil, for example, arose out of the emergence of nationalist movements that merged with a recuperation of African cultural traditions in attempts to reclaim and re-insert Afro-Latinos into the nation-building project (Chasteen, 2004; Scher, 2003; Trigo, 1993).

Race and ethnicity have been tied very closely to carnival celebrations in Uruguay and Argentina from the beginning of the arrival of Africans and Europeans during the colonial periods. Most popular press accounts of carnival in the early twentieth century continue to acknowledge a visible organized Afro-Uruguayan population who continue to play active roles in carnival there. In Buenos Aires, in contrast, there is very little acknowledgement of a similar phenomenon. If current-day carnival practices are strongly linked to the areas in the Americas which had strong African-descent populations, could the answer as to why Uruguay was able to maintain its carnival tradition be found within this segment of its population? Abril Trigo argues,

'Despite the fact that Uruguay is known for its Europeanized culture, the black minority, which constitutes about two percent of the population, has exercised an influence on urban popular culture that greatly exceeds its numbers' (Trigo, 1993, p. 716).

George Reid Andrews has excavated the history of Afro-Argentines, calling into question the degree to which this segment of the population 'disappeared' at the beginning of the twentieth century, and more recent scholarship is uncovering the various organizational societies that Afro-Argentines participated in during the early twentieth century (López & Dominga Molina, 2001, pp. 332–347). Andrews illustrates how the promotion of whiteness was intricately linked to the erasure of blackness (Andrews, 2004). For example, he shows how most cultural expressions linked to an African cultural tradition or to its African descendant purveyors were outlawed and heavily repressed. These laws had greatest effects where individual Latin American countries had the largest Afro-American populations, but they were, nonetheless, felt everywhere and contributed to the erasing of not only Afro-Latin American cultural traditions but also to this group's visibility in society. This erasure of blackness can be seen most visibly in carnival celebrations, since they have garnered a great deal of attention from legislators as well as the popular press, as mentioned at the outset of this article. The transformation of carnival in Buenos Aires from a celebration that reflected Afro-Argentine traits into one that reflected European ones was strong enough and fast enough to be captured in Domingo F. Sarmiento's description of how carnival in Buenos Aires was no longer celebrated with the display of flags representing African nations. Instead, the flags of the nations of the European immigrants had replaced them (Andrews, 2004, p. 106).

Between Buenos Aires and Montevideo, the clearest signs of the maintenance of African culture and populations were seen in Uruguay, where it is estimated that two percent of the population has African heritage – double that usually estimated for Argentina at the same time period. While mainstream newspapers very rarely commented on the role of Afro-Argentines in Buenos Aires during carnival, their presence in Uruguay was noted regularly and thoroughly, suggesting a prominent role in carnival celebrations for both Afro-Uruguayans and black face *Lúbolo* groups. Andrews discusses relationship between these minstrelsy groups and social constructions of race in Uruguay's carnival in the 1880s (Andrews, 2007, pp. 704–709). *Comparsas* which had evolved from African-based *candombes*, retained the participation of many Afro-Uruguayans, although the majority of them were composed of people of European descent. It was not unheard of to see mixed-race *comparsas* either. Newspaper ads continue to describe *comparsas* by their ethnic components, often noting when black *comparsas* were particularly 'organized and professional', probably code for those *comparsas* which had responded to the dominant cultural norms of the mainstream political and social classes.

It would be an overstatement to argue that carnival in Argentina did not retain Afro-Argentine elements only it was found more strikingly in the increasingly popular tango music rather than in carnival's *candombes* or *comparsas*. Many scholars have traced the roots of tango back to the *candombe* or slave dances of the early colonial era (Castro, 1991; Rossi, 1926; Chasteen, 2004). The dances of slaves transformed over time and the new dance form that they practised was originally referred to as the *tambo* then the *candombe* and eventually the tango. Chasteen suggests that the tango arose out of the practice of elite Argentine youth imitating the dance of Afro-Argentines (Chasteen, 2000). At the end of the nineteenth century, for example, as the Afro-Argentine population was decreasing at a rapid rate, it became common for the Argentine elite to dress up in blackface for carnival celebrations. An article in 1902 *Caras y Caretas*, 'El Carnaval Antiguo, Los Candomberos' (cited in Anderson, 2005), traces the rise of the elite youth, or *niños bien*, who imitated Afro-Argentine *candombes* by dressing up in blackface to 1870. An Afro-Argentine woman, wife of one of the organizers of the Benguela nation, interviewed for the article, reminisced that in the carnival of yore, when African nations participated, 'even the stones danced'. This all changed, she said, after young white men took to imitating the speech and dance of Afro-Argentines. A notable difference between Argentina and Uruguay was the fact that Afro-Uruguayan *candombes* continued to participate in carnival in Montevideo, but their counterparts disappeared from carnival in Buenos Aires (Andrews, 2007, p. 694).

The degree to which a stronger Afro-Uruguayan cultural tradition preserved carnival in Montevideo cannot be known. It does seem reasonable to suggest, however, that the same elements that fomented the maintenance of carnival there, notably the inclusion of the city's many diverse neighborhoods in primary participatory roles, also allowed the city's Afro-Uruguayan population to maintain and continue a strong organizational culture that, in turn, was able to preserve its own unique traditions and components which could then expand as political and social spaces embraced them. The disappearance of carnival in Buenos Aires served as yet another example of how the Afro-Argentine participation and presence in this city were forgotten.

Conclusion

Carnival in Buenos Aires, Argentina and Montevideo, Uruguay, experienced similar transformations as a result of the era of modernization, which stressed the ruling elite's desire to maintain public order and to foment consumerism in popular celebrations. Yet carnival in each city experienced these processes very differently. Buenos Aires' carnival shifted from city streets into theaters, and Montevideo's outdoor carnival continued to thrive. The two carnivals were clearly linked by the movement of the middle and upper classes from Buenos Aires to Montevideo, as a result of the commercialization of leisure time, and the development of a tourism industry in Montevideo.

Buenos Aires' carnival faded away due to a variety of reasons. The strong ideological position of its elites to propel its citizens on a strictly defined path of what it meant to be civilized resulted in a centuries-long attempt to control carnival revelry. Over time, city officials and the police force chipped away at the celebration's outdoor and spontaneous expressions. It is likely that the most consistent form of popular form that carnival took in Buenos Aires – throwing water and other objects onto passers-by – annoyed more citizens than it entertained. The rapid growth of the city which consistently outpaced that of its eastern neighbor fourfold is perhaps the most compelling reason to explain the demise of carnival in Buenos Aires. As a result of modernization, carnival began to fit in less well with the characteristics of an urban center and more and more with the traits of a 'gran aldea' (or 'large village' the descriptor most used to define Buenos Aires' up until about 1880), qualities that Montevideo retained. Academics have argued that the elite over-regulated carnival in

Buenos Aires but it is more accurate to argue that the emerging middle classes also rejected many aspects of outdoor carnival celebrations that hampered business as well as threatening personal security. In Montevideo, government regulation, rather than strangling popular festivities, served as a means by which the city's inhabitants and regulators could negotiate a functional form of carnival. The commercialization of leisure time and the way in which Montevideo responded to the new opportunities that tourism presented also promoted the retention and expansion of carnival activities here.

Notes

1. I reviewed issues of *Caras y Caretas*, *P.B.T.*, and *Fray Mocho*, middle to upper-middle class magazines, which included annual coverage of carnival, between 1900 and 1915.
2. Matinees arose in many cities about the same time due to the conflation of the rise of commercialization of cultural activities and the increase in leisure time afforded to urban workers. Matinees were especially prominent in Buenos Aires after 1900 where theaters catered to children by handing out candies during afternoon performances.
3. Almost every issue of *P.B.T.* in the summer of 1905, for example, included photos of Montevideo celebrations, most including the participation of Argentines. *Fray Mocho* and *Caras and Caretas* between 1910 and 1915 regularly covered events in Montevideo including, carnival celebrations for its Argentine audience.
4. Until recently, many more book-length studies have been devoted to carnival in Montevideo than to its counterpart in Buenos Aires. The most important books related to carnival's development in Uruguay include Alfaro (1991), Neto (1967), Remedi (2004), and Plácido (1966). Oscar Chamosa has noted that carnival has generally been a side note to Argentine historiography (Chamosa, 2003a).

Newspapers and magazines

Caras y Caretas
The Economist
Fray Mocho
La Gaceta Músical
La Ilustración Sud Americana
La Razón (Buenos Aires)
La Razón (Montevideo)
P.B.T.
La Prensa
La Tribuna Popular

References

Aching, G. (2002). *Masking and power: Carnival and popular culture in the Caribbean.* Minneapolis: University of Minnesota Press.
Alfaro, M. (1991). *Carnaval: Una historia social de Montevideo desde la perspectiva de la fiesta* [Carnival: A social history of Montevideo from the festival's perspective]. Montevideo: Trilce.
Anderson, P. (2005). Todo el año es carnaval [Carnival all year long]: Unmasking the carnival in Buenos Aires, 1810–1920. *Studies in Latin American Popular Culture, 24*, 37–56.
Andrews, G.R. (1980). *The Afro-Argentines of Buenos Aires, 1800–1900.* Madison: University of Wisconsin.

Andrews, G.R. (2004). *Afro Latin America, 1800–2000*. Oxford and New York: Oxford University Press.
Andrews, G.R. (2007). Remembering Africa, Inventing Uruguay: Sociedades de Negros in the Montevideo Carnival, 1865–1930. *Hispanic American Historical Review*, *87*(4), 693–726.
Bakhtin, M. (1984). *Rabelais and his world* (Hélène Iswolsky, Trans.). Bloomington: Indiana University Press. (Original work published 1968)
Bettelheim, J., & Nunley, J.W. (1988). *Caribbean festival arts: Each and every bit of difference*. St. Louis, MO: University of Washington Press.
Buenos Aires. (1881). *Memoria de la Policia* [Police Annual Report].
Buenos Aires. (1887). *Repertorio de Policia* [Minutes of the Buenos Aires Police].
Buenos Aires. (1900). *Sesiones del Concejo Deliberante* [Minutes of the Buenos Aires City Council].
Buenos Aires. (1901). *Memoria de la Ciudad de Buenos Aires* [City of Buenos Aires Annual Report].
Buenos Aires. (1923). *Dispociónes de la Policia, 1880–1923* [Police Regulations, 1880–1923].
Castro, Donald S. (1991). *The Argentine tango as social history, 1880–1955: The soul of the people*. Lewiston: E. Mellen Press.
Chamosa, O. (2003a). Lúbolos, tenorios y Moreiras: reforma liberal y cultura popular en el carnaval de Buenos Aires de la segunda mitad del siglo XIX [Lúbolos, tenorios and Moreiras: Liberal reform and popular culture in Buenos Aires' carnival in the second half of the nineteenth century]. In H.S.a.A.R. Lettieri (Ed.), *La vida política en la Argentina del siglo XIX: Armas, votos y voces* [The political life in Argentina during the nineteenth century: Weapons, votes and voices] (pp. 115–135). Buenos Aires: Fondo de Cultura Económica.
Chamosa, O. (2003b). 'To honor the ashes of their forebears': The rise and crisis of African nations in the post-independence state of Buenos Aires, 1820–1860. *The Americas*, *59*(3), 347–378.
Chasteen, J.C. (1996). The prehistory of samba: Carnival dancing in Río de Janeiro, 1840–1917. *Journal of Latin American Studies*, *28*, 29–47.
Chasteen, J.C. (2000). Black kings, blackface carnival, and nineteenth-century origins of the tango. In W. Beezley & L. Curcio (Eds), *Latin American popular culture: An introduction* (pp. 43–60). Wilmington, Delaware: Scholarly Resources.
Chasteen, J.C. (2004). *National rhythms, African roots: The deep history of Latin American popular dance*. Albuquerque: University of New Mexico.
Clemens, E.J.M. (1886). *La Plata countries of South America*. Philadelphia: J.B. Lippincott Company.
Crook, L.N. (1993). Black consciousness, samba reggae, and the re-Africanization of Bahian carnival music in Brazil. *World of Music*, *5*(2), 90–108.
Da Matta, R. (1991). *Carnivals, rogues, and heroes: An interpretation of the Brazilian dilemma* (J. Drury, Trans.). Indianapolis: University of Notre Dame Press.
Dunn, C. (1999). Afro-Bahian carnival: A stage for protest. *Afro-Hispanic Review*, *11*, 11–20.
Ergood, B. (1991). Os blocos de Santa Rita de Sapucai: Carnival clubs in a small Brazilian city, more than culture producers. *Studies in Latin American Popular Culture*, *10*, 141–158.
Graham, S.L. (1992). *House and street: The domestic world of servants and masters in nineteenth-century Rio de Janeiro*. Austin: University of Texas Press.
Koebel, W.H. (1920). *The great south land; the river Plate and southern Brazil of to-day*. New York: Dodd, Mead and company.
López, M.L., & Dominga Molina, L. (2001). Afro Argentineans: 'Forgotten' and disappeared – yet still present. In S.S. Walker (Ed.), *African roots, American traditions: Africa in the creation of the Americas* (pp. 332–347). Lanham, MD: Rowman and Littlefield.
Lynch, J. (1981). *Argentine dictator: Juan Manuel De Rosas, 1829–1852*. Oxford and New York: Oxford University Press.
McCleary, K. (2002). *Culture and commerce: An urban history of theater in Buenos Aires, 1890–1920*. Unpublished Dissertation, University of California, Los Angeles, Los Angeles.
Moya, J. (1998). *Cousins and strangers: Spanish immigrants in Buenos Aires, 1850–1930*. Berkeley and Los Angeles: University of California Press.
Needell, J. (1995). Río de Janeiro and Buenos Aires: Public space and public consciousness in fin-de-siecle Latin America. *Comparative Studies in Society and History*, *37*(3), 519–541.

Neto, P.d.C. (1967). *El carnaval de Montevideo; folklore, historia, sociología*. Sevilla: Facultad de Filosofía y Letras, Universidad, de Sevilla, España.

Panizza, F. (1994). *Uruguay, batllismo y después: Pacheco, militares y tupamaros en la crisis del Uruguay batllista* [Uruguay, Batllismo and after: Pacheco, soldiers and Tupumaros in the crisis of Uruguayan Batllismo]. Montevideo: Ediciones de la Banda Oriental.

Panizza, F. (1997). Late Institutionalisation and Early Modernisation: The Emergence of Uruguay's Liberal Democratic Political Order. *Journal of Latin American Studies, 29*(3), 667–691.

Plácido, Antonio D. (1966). *Carnaval: Evocación de Montevideo en la historia y la tradicíon* [The tradition and history of carnival in Montevideo]. Montevideo: Letras.

Puccia, E.H. (1974). *Breve historia del carnaval porteño* [Brief history of carnival in Buenos Aires]. Buenos Aires: Municipalidad de la Ciudad de Buenos Aires.

Puccia, E.H. (2000). *Historia del carnaval porteño* [History of carnival in Buenos Aires]. Buenos Aires: Academia Porteña del Lunfardo.

Remedi, G. (2004). *Carnival theater: Uruguay's popular performers and national culture*. Minneapolis: University of Minnesota Press.

Rilla, G.C.a.J. (2005). *História contemporánea del Uruguay: De la colonia al siglo XXI* [Contemporary history of Uruguay: from the colonial era to the twenty-first century]. Montevideo: Editorial Fin de Siglo.

Rosenthal, A. (1996). Dangerous streets: Trolleys, organized labor and the reorganization of public space in Montevideo, Uruguay. In R. Pineo & J. Baer (Eds), *Cities of hope: People, protests and progress in urbanizing Latin America, 1870–1930* (pp. 30–52). Wilmington, DE: Scholarly Resources.

Ross, G. (1916). *Argentina and Uruguay*. New York: The Macmillan Company.

Rossi, V. (1926). *Cosas de negros: Cosas de Negros: Los orígenes del tango y otros aportes al folklore rioplatense. Rectificaciones históricas* [Things of blacks: The origin of the tango and other cultural markers of the River Plate region]. Buenos Aires: Taurus.

Sarmiento, D.F. (2003). *Facundo: Civilization and Barbarism* (K. Ross, Trans.). Berkeley and London: University of California Press. (Original work published 1845)

Scher, P.W. (2003). *Carnival and the formation of a Caribbean transnation*. Gainesville: University of Florida.

Seigel, M. (2000). Cocoliche's romp: Fun with nationalism at Argentina's carnaval. *The Drama Review, 24*(2), 56–84.

Stallybrass, P., & White, A. (1986). *The poetics and politics of transgression*. Ithaca, NY: Cornell University Press.

Trigo, A. (1993). Candombe and the reterritorialization of culture. *Callaloo, 16*(3), 716–728.

Xosé de Enríquez y Vidal. (2004). *Momo Encandenado: Crónica del carnaval en los años de la dictadura (1972–1985)* [Momo unchained: Chronicle of carnival during the years of the dictatorship]. Montevideo: Ediciones Cruz del Sur.

Incremental art: negotiating the route of London's Notting Hill Carnival

Lesley Ferris

Department of Theatre, Ohio State University, Columbus, Ohio USA

This essay provides a brief overview of London's carnival and its beginnings in the late 1950s. Claudia Jones committed herself to both the culture and political underpinning of Caribbean carnival when she founded the event. London's West Indian community embraced carnival as an important source of celebration and cultural identity in the face of racist intimidation in Britain. The essay explores various difficulties that black British artists face gaining recognition, particularly those who work in carnival. Clary Salandy, the artistic leader of Mahogany Arts Ltd., in London, is considered one of London's leading, vanguard carnival designers. Her creation of a mas camp and her focus on community development to pursue carnival as a collective enterprise is examined. Salandy insists that 'Carnival is art' and her work exemplifies the tension between her desire to establish carnival as a viable art form in Britain and the social, political and cultural barriers faced by carnival artists today.

Introduction

Today London's Notting Hill Carnival is considered the largest street theatre event in Europe. It is a calendrical carnival, taking place yearly on the Bank Holiday weekend (the last Sunday/Monday in August). The outdoor street procession, food stalls and the numerous static sound systems that are nestled within the carnival route that circulates through Notting Hill in west London are a testament to the energy and commitment of Claudia Jones, a political activist and writer, who founded the Caribbean Carnival Committee in November 1958. Jones, who also launched the first black newspaper in Britain, *The West Indian Gazette*, organized the first carnival in response to the racist intimidation and violence from white youths against West Indians that escalated in the summer of 1958.

London's first Caribbean Carnival took place on 30 January 1959 in St. Pancras Town Hall. The program printed for this first carnival included a manifesto-like statement from Jones asking the West Indian community to help keep this carnival an annual event (Sherwood, 1999, p. 156) and it continued to take place until the sudden and unexpected death of Claudia Jones in December 1964 (Boyce Davies, 2007, p. 182). Following Jones' death, the West Indian celebrations continued, but outdoors, relocated to Notting Hill where the largest Trinidadian population settled.

In 1973 the first costume bands went on the road bringing the Trinidad tradition of carnival design to London. A band in this sense is a group of people who come together to take part in carnival, or to 'play mas'. The costumes for a mas band are designed by a carnival artist and usually have a theme or a story. While the Caribbean term 'mas' derives from the word 'masquerade', the term is used in a much different way, as explained here:

> Those who go on the road in a costume band in carnival are called mas players. However, simply wearing a costume is not mas. It becomes mas when the player plays it, connecting to its meaning from inside him/herself and giving that character or thematic aspect full life on the street. European traditions of masquerade usually involve the wearing of the face mask. In Caribbean mas, a mask may occasionally be worn, but more important is how the character is played or the theme is animated and given meaning and purpose by the player. (Ferris & Tompsett, 2007, p. 47)

The face mask is not the only distinction between Caribbean and European carnival. A crucial difference is Caribbean carnival's indelible link to emancipation. In 1783 the French colonial powers introduced Mardi Gras masquerade balls into Trinidad. When Britain seized Trinidad in 1797, the existing French plantation system, and thus the European Mardi Gras, continued. The abolition of slavery in the British colonies in 1834 brought the now-freed Africans onto the streets to celebrate and by 1838 the French Mardi Gras was transformed to an Afro-Caribbean Carnival. Carnival developed in the Caribbean over the following decades as a vibrant and at times embattled occasion of resistance art. Despite the advent of abolition, the British colonial powers made every effort to control and contain the newly configured carnival. With independence from Britain in 1962, Eric Williams, the first prime minister of Trinidad and Tobago, recognized carnival as the national art form.

Carnival's arrival in London developed as part of the British government's post-war initiative to rebuild the damaged country with help from its colonies. In 1948 the first ship, the SS Empire Windrush, docked in Britain containing hundreds of West Indians, including the famous calypsonian, Lord Kitchener, who was captured on newsreel footage singing his calypso 'London is the Place for Me'.

With the growth of London's carnival (current attendance figurers are running at well over one and a half million), the aspirations of its talented artists who make this visual spectacle possible have intensified. While carnival artists from Trinidad, such as Peter Minshall and Wayne Berkeley, are considered major artists in their own right, such recognition has been slow in coming to the British-based artists.[1] As Kwesi Owusu says in his 1986 study *The Struggle for Black Arts in Britain*:

> There is in western art a central artistic concern for the creation of illusory space which is quite different from that occupied by we mortals. Illusory space requires artificial containment to separate it from real space. Frames and proscenium arches accomplish this containment. Galleries, museums and other institutional spaces separate visual art from performing art, books from film, music from poetry. African and Asian carvings and masquerade, on the other hand, share with mortals the real space of the physical world. Masquerades, which combine visual and performing arts, occur in the streets... The physical and social landscape is the immediate canvas of creativity. (1986, p. 81)

Owusu's commentary points to several issues surrounding traditional Western notions of art. Carnival art is not made to be contained in an 'illusory' building-based space, indeed its unique contribution is that it is framed by the street, a social landscape that constantly connects the carnivalist to the every day world. It is the very interplay between street and artist, between the west London houses and dancing mas players, that give carnival its force and power. Not only is the street an experiential space, it is redolent with history. Adela Ruth Tompsett explains the importance of the geography:

> [T]he processional route at Notting Hill resonates with a particular and pertinent history. It is a microcosm of that macrocosm that is a shared history between colonizer and colonized. From the industrial canal behind Kensal Road and the old factory buildings and working man's club in the north of the area, down to the fine streets of Arundel Gardens, Kensington Park and Westbourne Grove, in the south, with their handsome white stucco fronted houses built with the wealth of the empire, the route reflects every level of empire activity from the making of wealth of plantation and factory to the displaying of it in the grand houses. (2005, p. 49)

Carol Boyce Davies concurs: 'Carnivals, in the African diaspora tradition, demonstrate the joy that its people experience in "taking space"' (2007, p. 167). Thus carnival art speaks to geographic freedom, and it is this metaphorical speaking, found in the mas players' procession, that conveys meaning to the work.

Since beginning over forty year ago, London's Notting Hill Carnival has survived, and even thrived, despite the fact that it continues to be 'a contested event'. As Abner Cohen states, 'every major carnival is precariously poised between the affirmation of the established order and its rejection' (Cohen, 1993, p. 3). The 'affirmation' from the powers that be, however, rarely happened during its early and middle years. Even today the local Borough of Kensington and Chelsea (currently the only Conservative controlled borough in London), where Notting Hill is located, strongly opposes its presence, despite the support carnival receives from other prominent government agencies. While conservatives in Kensington and Chelsea worked to get the event moved to another location, the carnival community fervently resisted the notion of relocation. As Cohen explains:

> Notting Hill as a place held a special symbolic significance for the West Indians who regarded it as, in the words of one leader, the nearest thing they had to a liberated territory – implicit reference to battles they had fought there against white racists in 1958 and against the police in 1970 and 1976. The carnival continued to exist, its greatest political achievement being that it had survived at all, in the face of formidable opposition and pressures operating to subvert it all the time. (1993, p. 3)

As the years passed since London carnival's beginning, the Notting Hill area, like many other impoverished areas in the center of the city, became sites of gentrification, as developers saw the opportunity to convert the old, dilapidated houses into modern flats that could only be sold to prosperous homeowners. This only intensified the local opposition to the event as many of the new homeowners joined the local government in demanding that the event be banned from Notting Hill. It is perhaps no accident that some of the most intense pressure against the carnival came during Margaret Thatcher's 11 year reign as a conservative Prime Minister (1979–1990). One police officer with the Metropolitan police during the Thatcher era states:

> Thatcher let it be known to us, the police, that we could do anything to keep in control the enemy within. They were blacks, trade unionists, and people who did not agree with her views. I know that when I was on duty in Notting Hill Gate, I would go for the blacks more than I should have done, but you get into a kind of state, like you are in the army and the enemy is the enemy. No wonder the blacks never trusted us. They would be idiots to. (Alibhai-Brown, 2001, p. 81)

Tompsett summarizes this situation as follows: 'In the 1970s and 80s, carnival in London was nationally – through the representation of it in the media – seen as a potentially disordered event. The police, institutionally racist, saw carnival and especially the participation of black youth as trouble-loaded'. Thus London's carnival faced two major hurdles: first its representation in the mainstream press as violent and lacking control combined with a deeply imbedded belief that art could not possibly happen outside of conventional building-based spaces, those sites of 'artificial containment' as evoked earlier by Kwesi Owusu. Tompsett continues, 'Carnival and "art" were not seen as in any way connected by the authorities' (2006).

A further argument used against carnival's artistic claims is that carnival takes place only once a year. Art making, so the conventional paradigm maintains, is not seasonal, not confined to a single two-day event. Furthermore, the carnival artistic process is collective, a group activity, which diverges from the Western notion of the solitary artistic genius. While lead designers provide the concepts and themes, producing and making hundreds of costumes is a team effort. Contrary to the limited conventional point of view, is the position that Cohen articulates, 'Mas making is an art form in its own right that involves knowledge, skills, imagination and creativity' (1993, p. 101). Added to this idea of carnival as an 'art form in its own right' is the political consequence of carnival. Let us return to the founding mother of London's carnival. Carole Boyce Davies in her recent study of Claudia Jones states:

> The intellectual, cultural and political work of Claudia Jones, in my estimation, offers one of the best models of African diaspora activism available. For Jones, who emerged as a leading member of the developing London Caribbean community in the 1950s and early 1960s, carnival carried, in its cultural practice, resistance to Euro-American bourgeois aesthetics, imperialism and cultural hegemony, and political and racial oppression. (2007, p. 167)

Much has been written about the many contradictions and paradoxes that are seen to define carnival in a variety of ways. Milla Cozart Riggio considers the potential limitations of carnival's binary logic in her introductory essay 'Time Out or Time In? The Urban Dialectic of Carnival': 'Carnival's primary source of energy – and this has a lot to do with its flexibility – is located neither fully in inversion nor in affirmation but in the tension between subverting and affirming, or, put another way, in its dialectic between civilized respectability and vagabondage' (2004, p. 19). A particular tension that is specific to London's carnival is that between the staging of the event itself (and as we have seen above certain powerful, local resistances to it) and the occasionally sanctioned support of the event. Errol Hill notes that the British Commission on Racial Equality wrote to the carnival organizers in London in 1994 with the following: 'Against the ominous backdrop of ethnic conflict and rising race hate politics in much of Europe, carnival stands out like a joyous beacon of hope and unity' (1997, pp. xxiii–xxiv). So here is a form of government endorsement and affirmation that is in tension with positions held by the local borough and police.

Carnival artist in London: the case of Clary Salandy

While recognition for London's carnival artists has been slow and incremental, taking place in fits and starts over the last decade or so, these artists see themselves as part of cultural tradition of resistance, working within an established art form (albeit one that is little recognized as such). And like any artist, carnival artists experiment, develop ideas, and push themselves into new arenas. In the following, I offer a case study of a London carnival artist, Clary Salandy, who has been working as one for over twenty years. According to Keith Nurse, there are between 25 to 30 carnival artists based in London, and others located in various regional cities where British carnival takes place, such as Leeds, Bristol, Birmingham, Manchester, Liverpool, Luton and Leicester (2004, pp. 245–254). Salandy is acknowledged as one of the artistic leaders and in 2007 she was one of four featured carnival artists in an exhibition at City Hall, London.[2] Of the four featured artists, she is the only one who continues to take part in Notting Hill Carnival with a fully designed costume band.

Mahogany Arts Ltd., a mas(querade) camp founded in 1989 by Clary Salandy and Michael Ramdeen, has a storefront on the High Street in Harlesden in northwest London. The name *mahogany* was chosen for the company because its name means 'strong, hard, long lasting wood. It is also a wood that can be carved and shaped, thus mahogany is a natural material that represents our vision, our art' (Salandy, personal communication, 29 November 2005). Salandy considered naming the company Visage, as one of her artistic interests is a fascination with the human face. But the name Mahogany seemed at the time to provide a more significant hallmark as it offered a sense of permanence, durability, and resilience, all qualities she hoped would soon be attributed to the company. Unlike the majority of carnival workshops in London, Mahogany has a permanent home, located in the London Borough of Brent. In addition to a large population of Indians from the subcontinent, Harlesden is known as the center of Brent's large Afro-Caribbean community. Also unlike other mas camps, Mahogany stays open all year round, catering to a range and variety of carnival design commissions, work that makes it possible to stay open and to employ a team of design assistants and technicians. The many commissions and invitations to present Mahogany's work at a variety of venues (from shopping mall openings to the Afro Asian games in India) are a means to an end, for the centerpiece of the work is playing mas at Notting Hill Carnival.

Salandy conceived and designed some of the most spectacular and award-winning carnival designs over the last decade. Her founding partner, Michael Ramdeen, serves as the technical director for Mahogany, working to create light-weight structures – such as back packs, wheeled carts that support enormous skirts, and fiber-glass fishing rods that support billowing silk fabric – practical structures worn by carnival performers that enhance a costume's kinetic quality.

Salandy says, 'Carnival is always telling you something. It's a language – if you can understand the language then you can read carnival' (Tompsett, 2002, p. 39). And it is this language that she pursues by pushing the boundaries of the art form while simultaneously proclaiming that carnival art *is* art. Her artistic vision is indelibly linked to her Trinidadian heritage. She explains, 'I feel the ancestral pull.

Before I was in this world, people were making carnival and they made me who I am today. We have to respect this and I start from this standpoint: that people died for this. We are paying tribute to the past' (Salandy, personal communication, 29 November 2005). Salandy explains the impact that reading Errol Hill's book, *The Trinidad Carnival: A Mandate for a National Theatre*, had on her. Hill's groundbreaking work, first published in 1972 and reissued in 1997, was one of the first scholarly works to articulate the troubled history of Trinidad's carnival while also celebrating its revolutionary nature (Funk, 2003). As Errol Hill states:

> [C]arnival had become a symbol of freedom for the broad mass of the population and not merely a season for frivolous enjoyment. It had a ritualistic significance, rooted in the experience of slavery and in the celebration of freedom from slavery. In this sense, carnival was no longer a European-inspired nature festival. Adopted by the Trinidad people it became a deeply meaningful anniversary of deliverance from the most hateful form of human bondage. The people would not be intimidated; they would observe carnival in the manner they deemed most appropriate. (1997, p. 21)

For Salandy carnival is, importantly, as Hill states, a 'symbol of freedom', a celebration of emancipation, a 'statement that we are here now, in this place and time, and we are not where we were in 1834' (Salandy, personal communication, 29 November 2005). The year that Salandy references – 1834 – is the year of Trinidad's emancipation from British rule and the end of slavery.

Carnival in London, like that in Trinidad, combines two cultural forms: music and the mas band, or costume band. The kind of music that is central to carnival's West Indian heritage is calypso and steel drum, or pan, and more recently soca. An added musical component that is totally unique to London is the sound systems that are set up within the boundaries of the parade route in Notting Hill. The sound systems, originating from Jamaica and thus linking that island to London's event in an indelible way, were introduced to carnival for the first time in 1975. Their success brought thousands of West Indian youth on the streets. While music in its multiple forms is part of what makes the event carnival, it is the playing of mas that is considered the heart of the carnival experience. While people can play mas on their own, by simply showing up in their own costume, 'a real masquerade is a group activity, which is planned and carefully organized and choreographed by artists' (Cohen, 1993, p. 3).

Mas bands are designed as a series of sections with each section broken into subsections. At the Notting Hill Carnival sections have anywhere from 5 to 30 performers in them; entire bands have between a total of 30 (considered a small band) to 300 players (considered a large band). In addition to the sections, there are usually single, or individual costumes and the two most extravagant and grand costumes: the king and the queen. A key element to the design is the conception of a costume that can embody movement and energy, a kinetic quality that generates much of the meaning of the performance. As Salandy states, 'I never see [the costume] without seeing where it's going or how it's animated – it always has inherent movement. When the player moves, the fabric, the structure almost breathes... It's a marriage between the performer and myself – it's my concept which he or she is bringing to life' (Tompsett, 2002, p. 41). Salandy clarifies that 'a costume is like a visual poem and it has to speak to everybody whether you are 2 or 88' (Salandy, 2006).

Salandy trained as a theatre designer so her designs tend to emphasize story over theme. Some tell a literal story, such as her award-winning *Carnival Messiah* (2000) in which she retells the Christian story through four twentieth century messiah-like figures chosen to serve as historical markers and international icons for peace: Nelson Mandela, Gandhi, Martin Luther King, Jr., and Mother Theresa. Intermingled in this design are traditional masks, certain characters that 'reappear in essentially the same type of costumes, though the dresses are made afresh each year, elaborated upon, and have a different design of basic colors' (Hill, 1997, p. 87). For *Carnival Messiah* Salandy drew on the traditional mask of the Devil as a narrative counterpoint to the other characters. Hill explains the particular significance of the Devil in carnival performance: 'Old religious doctrine held that music and dance were both inventions of the devil. Hence, no carnival could be considered complete without the representation of His Satanic Majesty in one or another of his many forms' (Hill, 1997, p. 88). Salandy's design plays on multiple devil figures: the multiple Dark Angels, the Serpent, and a single dancing Devil, representing historical battles against the forces of Christianity. This dancing Devil often interacted with solitary white Angels who roamed the final two sections of the *Carnival Messiah* band. These mock battles between good and evil took place with comic ferocity to the delight of the carnival

Figure 1. The dancing devil with one of the angels in *Carnival Messiah*, designed by Clary Salandy for Notting Hill Carnival 2000. (Photo: Lesley Ferris)

spectators (Figure 1). The Angel Gabriel, an individual costume with billowing wings, is another character, who with a section called Doves of Peace, served as an antidote to the Dark Angels and their underworld (Figure 2). For Salandy this mixing of contemporary concerns with historic carnival characters and the Christ story is a way of making social commentary through her art.

Carnival Messiah can be considered a watershed moment in Salandy's carnival design work, partly because of the boldness of her design and partly because of the multiple times aspects of this work have been staged as a theatre production. It has an atypical history as it began life as a theatre design by Salandy for a production also called *Carnival Messiah* conceived and directed by Geraldine Connor for the West Yorkshire Playhouse in Leeds in 1999. This used the music of Handel's *Messiah* and told the familiar story through West Indian carnival costumes. Salandy, kept some of the designs from the theatre production, and put in others, which included the major alteration of adding the four twentieth century 'messiahs' for Notting Hill in 2000. The theatre production has had additional restagings and revivals, the most recent being in autumn 2007.

Figure 2. 'The Angel Gabriel' surrounded by the 'Doves of Peace' in Mahogany's *Carnival Messiah* in 2000. (Photo: Lesley Ferris)

A particular design element employed by Salandy is the shield, which she used in multiple ways in *Carnival Messiah*. She explains her interest in this as follows:

> I love using the shield because it is a symbol of Africa. It's a symbol of war and I feel that in some ways I am a contemporary warrior. I'm holding on to this culture, protecting it, taking it forward and badgering the police as we go through and use the roads. In *Carnival Messiah* I called them 'shields of honour', they were lightweight, made from silk, and the face of Mandela was printed on the front in red. (2007, pp. 32–33)

In one particular section of this band, a group of mas players were dressed as Shango priestesses, they held the silk shields and danced through the streets (Figure 3). The lightweight shield as a source of protection is an ironic gesture, one that plays on carnival's ability to proffer multiple readings at once. The shield evokes 'Africa' as a symbol, a memory, or as Stuart Hall states as a '"translation moment" of the re-encounter with Afro-Caribbean traditions. The reason is not that Africa is a fixed anthropological reference . . . [but that] "Africa" is the signifier, the metaphor, for that dimension of our society and history that has been massively suppressed, systematically dishonoured and endlessly disavowed . . . Race remains, in spite of everything, the guilty secret, the hidden code, the unspeakable trauma, in the Caribbean. It is "Africa" that has made it "speakable", as a social and cultural condition of our existence' (Hall, 1999, p. 14). Salandy's designs over the years have engaged time and again with African themes and her work provides a contemporary example of an artist that makes 'speakable' Hall's concept of 'Africa'.

Over the next several years Salandy continued to design carnival bands that focused on African art and history such as *A Touch of Africa* (2001), *Out of This*

Figure 3. 'The Shango Priestesses' holding the 'shields of honour' for Mahogany's 2000 carnival design *Carnival Messiah*. (Photo: Lesley Ferris)

World (2002), *The Afro-Asian Experience* (2004), *Mosaic* (2006), and *Freedom Song* (2007). Her work created for 2005 carnival was a departure for her as it moved outside her previous methods of storytelling so evident in *Carnival Messiah*. Her 2005 band was titled *A Stretch of the Imagination*, and her design explored shape, line, form, volume, depth and scale. While one could say that this is what carnival designers do as a matter of course, this is commonly achieved in the context of a theme or story. With *A Stretch of the Imagination* Salandy eschewed storytelling in order to explore an abstract aesthetic. She sees this approach as an artist's right and duty to experiment, to push the boundaries of what is possible.

Salandy describes one of her self-imposed artistic challenges as an attempt 'to capture the essence of the thing' (Salandy, personal communication, 29 November 2005). In the 2004's mas band, the *Afro-Asian Experience*, she created a range of different animals: giraffes, zebras, elephants, peacocks. With contemporary technology she developed with her partner Michael Ramdeen a combination of lightweight fiberglass rods and thin foam which is molded and shaped to create spectacular, life-like creatures that are brought to life by the carnival performer. She talks about the ways in which she wanted to 'catch the essence of zebra-ness' in a way that utilizes the shape and line of the human body with her costumes. But with her 2005 work she moved away from representational figures to the abstract.

A Stretch of the Imagination had over six sections of between 15 to 20 players in each. The names of the sections are reminiscent of titles of abstract paintings one might see in a modern art museum: 'Volume in Black and Blue', 'Study in Gold and Orange', 'Kaleidoscope', 'Red and White Contrast' and 'Natural Form in Blue and White'. But unlike static two-dimensional art, these works are constantly moving, shifting and changing. In 'Volume in Black and Blue', the player wore a low-slung flat skirt that was six foot in diameter. Swirls of black and blue color made up the skirt. At a given moment, however, the player lifts up the flat skirt to transform it into a spiral shape that extends up and over the player's head. It is almost like taking a flat painting that is horizontal to the ground and converting it into a three-dimensional work with vertical as well as horizontal volume. 'Study in Gold and Orange' had over 20 players, young girls and women, each in a solid color of gold or orange, each with two fan-like shapes coming from the waist and pointing toward the ground. The players grabbed the top of the fan and lifted it to reveal a honeycomb-like shape, hidden within a flat fan that transformed to a three-dimensional web pointing skyward. One of the Adult Female Individual costumes, entitled 'All is Not What It Seems', had an enormous flat skirt, in swirling colors of blue, gold and mauve, which transformed into a three dimensional corkscrew. As the player danced through the streets of Notting Hill, she always drew applause and gasps of wonder when she lifted her skirt skyward, erasing her human shape and replacing it with a spiral (Figure 4). The Queen costume, entitled 'Out of the Blue', was a tight fitting vibrant turquoise blue dress with a fan-like spray at the bottom, mid-calf. The player lifted the fan shape and it opened to reveal a multi-colored interior, a web of geometric forms than transformed the Queen from slender shape to imposing winged creature (Figure 5).

Salandy did create one costume that was not abstract and that was her Carnival King. His name was 'In the Structure of My Mind', and against carnival tradition of a single king, she had two (Figure 6). This costume was an enormous human head, shaped out of lightweight fiberglass rods and covered with translucent material. One

Figure 4. Designed by Clary Salandy, *A Stretch of the Imagination* explored a variety of abstract shapes. Here the adult female costume, entitled 'All is Not What it Seems', began as a flat circular shaped skirt, which opens in a rising corkscrew motion making the player invisible. (Photo: Lesley Ferris)

costume was gold and one was silver and inside each head was the full body of the mas player himself. The costume was held in place by a cleverly designed backpack and the eyes of the giant head could open and shut, controlled by the player. Salandy has an artistic fascination with the shape, contour, and structure of the human face. All of her designs over the years have featured large faces in a variety of forms. Walking past her storefront studio, many of these large heads are showcased in the Mahogany windows. Salandy described her concept for 'In the Structure of My Mind' as follows:

> Like the brain is the engine room for the body, here the masquerader is the life force bringing the mas to life. The integrity of the structure is maintained through the simplicity of materials used and the translucent qualities of the piece allow us to see *the man behind the mas.* (2005)

The imposing large three-dimensional heads placed the human mind, the source of imagination and creativity, at the center of mas making. It appeared as if the head had conjured up and imagined all the abstract shapes, colors, and forms that followed the double-dancing kings down the streets.

Figure 5. The Queen costume, entitled 'Out of the Blue', for Mahogany's *A Stretch of the Imagination* for Notting Hill Carnival, 2005. (Photo: Lesley Ferris)

A central and well acknowledged characteristic of carnival is its ability to transform the every day into something out of this world. Carnival changes the streets of Notting Hill from urban, grubby thoroughfares to vibrant avenues of color, shapes, and forms moving and pulsating to a contagious musical beat. Salandy's costume art, as we have seem in her 2005 piece *A Stretch of the Imagination*, uses kinetic transformation as a crucial element of her artistic arsenal. Her experiments with shape, volume, texture are nothing without the player. As Tompsett explains, '[A] mas, or masquerade, is in itself neither a character nor a costume, but the combination of the two, given life by the player on the road. A costume is only a costume until the player plays mas' (Tompsett, 2005, p. 49).

Another aspect of transformation is Salandy's ability to make change in her local community. She works hard to develop community involvement that allows her to realize her artistic vision. She does see carnival as an 'art form that involves the community', and she is committed to what she describes as a 'social service dimension' to her work. Having a full time studio that is open year round, gives Mahogany a stability and presence in her neighborhood. Salandy states: 'I have a strong sense of being a role model as an artist and as a person who is focused and committed to what I do. It is extremely important that the young people who work here witness my overall sense of care and purpose' (Salandy, personal communication, 29 November 2005). Throughout the year Mahogany hosts over a dozen students in work experience placements that are organized by the educational authorities in the London Borough of Brent. In July 2004 a group of work experience students were given the task of making the foam shaped zebra heads that were featured in that year's design, *The Afro Asian Experience*. In addition to the

Figure 6. 'In the Structure of My Mind', one of the two King costumes, for Clary Salandy's *A stretch of the imagination*, Notting Hill Carnival, 29 August 2005. (Photo: Lesley Ferris)

temporary students, Mahogany employs up to a half a dozen design assistants and technicians, several who proved their worth as prior work-experience students and were invited to stay on in paid positions. Furthermore, Mahogany works with numerous volunteers from the local community, both young and old.

Salandy sees her community involvement as a two-way street. She realizes fully that her designs in their scale and number can only be realized as a result of a collective work process. In return, the people who work with her are able to take part in an artistic practice that is guaranteed an audience during London's Carnival. Salandy explains that 'what we produce must be excellent so that they [the children] can take part in something excellent in their lives' (Salandy, personal communication, 29 November 2005). It is through the local children that she has developed an artistic process that can be described as incremental. The system works as follows: Salandy designs certain costumes that have small pieces or sections that can be made by children of any age. Children see the large-scale works from previous carnivals and baulk at their size and construction challenge. Knowing this, Salandy demonstrates, often with the help of seasoned assistants, how to make one portion of the costume aimed at the skill level of the children. Then after all the smaller pieces are finished, the pieces are assembled, in many ways like a large puzzle coming to life. The children witness this process of carnival transformation from, for example, a small painted feather that they make which, when assembled, turns into a large imposing costume. They are willing to invest their time and effort in this process. While such collective artistic strategies are not new, Salandy plans her designs to take advantage of this process and to give a sense of ownership to her community volunteers.

Thus for Mahogany local children are vital participants in their studio work. Selina started to volunteer as a seven year old, and at age 16 (in 2006) she is one of the accomplished and committed volunteer teenagers. Salandy explains that 'people learn and grow up here. Carnival makes this happen' (Salandy, personal communication, 29 November 2005). When Mahogany receives invitations to special events outside of Notting Hill Carnival, such as the Queen's Jubilee, or when they are given a commission for an international event, such as the Afro Asian Games in India or the Millennial Celebrations in Singapore, it is these local children and young people who are able to go with Mahogany and perform. So these local children benefit in multiple ways from taking part in carnival making. Salandy is very aware that it is through the ever-changing community of young people that her ideas are realized. She has a fierce and passionate commitment to artistic values and does not see the children's work as lowering her artistic standards.

Salandy (2005) eschews recent trends in Trinidad and London where some carnival bands merely sport 'bangles, bikinis and beads', and where party culture supersedes the history, traditions, and art form of carnival. 'We will not do slap dash. We must treat the art with reverence. If we can't do it properly, we won't do it' (Salandy, personal communication, 29 November 2005).

'Bikinis and beads' refers to bands in which the costume design is more akin to a Las Vegas show girl style, than to the rich history of character, satire, and social and political comment found in traditional mas or the themed or story-telling bands of contemporary carnival. A particular mas band, known as Poison, began producing bands in Trinidad and Tobago several years ago and they are aptly named, says Trinidadian artist Christopher Crozier, because 'they replicate and spread like poison' (2005). Poison TT, as the original band is called, runs a sophisticated website, with flash design, music, and a panoply of advertising. It has a gallery of photos of the bikini type costumes, many with feathers protruding, fan-like from the backs of the women who wear these revealing outfits. The concept for the 2006 band was 'Shangri-La'. The introductory text on the Poison TT website is a worst-case example of over-the-top advertisement verbage: 'Welcome to Shangri-La, a place of myth and legend where every desire will unfold before your eyes'. And 'An All-Inclusive Dream Experience! We Make Your Dreams a Reality!!' The website is full of slick sales lingo such as 'We deliver a well-packaged product to the market place'. And it provides prices. For a standard female costume it will cost a carnival participant TT$1950, or US$315, or £175.[3]

This upfront, commercialized marketing version of carnival came to London in 2003 when Poison UK was founded, a direct 'spin-off' of the Trinidad and Tobago model. Hosting a well-designed website it also replicates the brash media-speak style of the original: 'Poison UK is all about the culture and lifestyle of the Caribbean which we believe can be mirrored here in the United Kingdom, and we would like to include you in this exclusive experience'. Under a section on the website entitled 'Services', Poison UK extols the authenticity of their Caribbean lifestyle and culture while simultaneously making it available as a commodity to anyone who would like to join their ranks. And like Poison TT this London version is all about the social life: 'In recent years the main focus has been placed on 'having fun' and in beautiful intricate costumes' (Poison UK, 2006). This shift to commercialized pleasure is for Salandy, and other London carnival artists and players, a worrying trend.

In contrast to this model of carnival entrepreneurship is Salandy's deeply embedded belief that through carnival 'we are paying tribute to the past'. A common practice for mas bands is to host a launch party of the costume designs some months before carnival. While Mahogany follows this tradition, Salandy makes it clear that Mahogany's work is *not* about the party. She is saddened by what she sees as a growing focus on the revelry and she makes sure there is no party culture – in its negative sense – in the Mahogany studio. She believes that 'the whole point of carnival is the making', giving the opportunity to the young people to see how their contribution, their piece of the larger artistic puzzle, comes together on the road. She will 'not sacrifice the art for the party' (Salandy, personal communication, 29 November 2005). And she does not want to sell her carnival band costumes. She feels that participants have a different attitude when they buy their costumes from the carnival artist, an attitude that works against the collective, community spirit she wants to instill in her band members. The Mahogany studio is available for all to come and work, to help make the 200 plus costumes for London's Notting Hill Carnival. But they are not for sale in the way that Poison TT and Poison UK sell their costumes.

Some might find Salandy's position on this a contradiction since she does sell her work in terms of specific commercial commissions. The commissions vary widely from creating chicken costumes for a Greenpeace demonstration to creating flower costumes for Nice's festival of flowers to opening a Grand Prix Racing event at Brands Hatch (south of London) to designing costumes for holiday celebrations for Selfridge's, an Oxford Street department store. Her rationale is that it is these commissions that make it possible for her to run her studio full time, to provide work placements for numerous students, to offer her local community in Harlesden a space where they can be part of a collective artistic effort, and to make it possible to work as a full time artist.

The demands of working as a full time artist cannot help but be intermingled with a desire for an acknowledgement of this position. For a carnival artist, the difficulty of gaining due recognition in a wider cultural context is daunting, despite a decided shift in perception on the part of various establishment institutions. In 1986 when Kwesi Owusu published his book *The Struggle for Black Arts in Britain* he summarized the many roadblocks black artists faced from lack of subsidized funding to a ghetto mentality that sees black art as 'ethnically other', to reducing its status by referring to it as 'popular culture' or 'folk art'. London's Notting Hill Carnival, with its history of negative media coverage, was viewed, in Cecil Gutzmore's word, as 'the domain of threatening culture' (in Owusu, 1986, p. 8). Since then there are a variety of positive moves to support and recognize carnival as an art form. In 1999 carnival artists were invited to perform at the Millennium Dome celebrations in front of the royal family and the prime minister. They were also the leading group in the Queen's Jubilee celebrations in 2003. During 2005 carnival artists were actively involved with

London's mayor and City Hall in presenting examples of their work for the Olympic selection committee. Many felt that London's successful bid for the 2012 Olympics was enhanced by the carnival art, which demonstrated a vibrant artistic diversity in the nation's capital. But there is still a long way to go.

It might be worth considering a particular incident that further demonstrates this point about acknowledging the value of the carnival artist. Salandy was invited by the Museum of London to present a family workshop on a Sunday in October 2005. With a laptop presentation she showed multiple images of her designs, had clips of videos to show the 'on the road' experience, and she brought several Mahogany helpers that were wearing costumes. The publicity office at the Museum did not publicize the presentation and it was difficult to even find mention of it on their website. It was attended by about 20 enthusiastic adults and their children, who just happened to be in the Museum at that time and heard a loudspeaker announcement inviting them to attend. So while offering the workshop is a promising educational opportunity for the museum, why relegate it to the margins in this apparently thoughtless fashion? This particular incident is not an isolated one. Many museums and art galleries in Britain tend to 'use' carnival art (if they choose to include it in any way) as an example of their commitment to diversity, but as we see here such commitment has an unfortunate 'window-dressing' feel to such inclusion. (For another example of the ways in which various institutions marginalize or erase carnival artists see my essay on the challenges of archiving and preserving carnival art, particularly the section entitled 'The Challenge of Research on Carnival: The Anonymous Artist', see Ferris, 2009, pp. 132–133.)

So while the process of carnival art is incremental, so to is the gradual and troubled route to recognizing carnival art and artists. But it does not stop Salandy from putting imaginative and bold artistic statements on the road at Notting Hill, work that is connected to the 'ancestral pull' that informs her work. Salandy described her original impetus for focusing her work on carnival as follows:

> My original motivation for becoming involved in carnival was my realisation that in England carnival was considered trivial and kitsch. These words came from my art teacher. How do I convince her that she is wrong? It was not trivial or kitsch to me. It was something that was deeply rooted in my culture. It made me the kind of person I am. (2007, p. 29)

Salandy was born in Trinidad and as a young adult she came to London to study theatre design. She continues:

> When I first came here, I was scared because carnival had such a bad reputation. When my teacher said carnival was kitsch, the only thing I could do was to ditch the fear and get in there and change it. You can only change it by going in and making it better. I believe the better the art is, the less people can ignore it. Carnival is art. (p. 33)

It is interesting to note here that Peter Minshall, the Trinidad carnival artist, also experienced a similar crisis of conscience in relation to his own activity as a carnival artist. Minshall explains his position: 'I wanted to give my island something of itself. My new-found understanding had banished my own self-contempt, and my mission became to share this understanding. To eliminate the self-contempt from all of us. To demonstrate that we have every reason to be secure, confident, and proud. And so, I became a "mas" man' (in Aching, 2002, p. 86).

Becoming a 'mas' man' or in the case of Clary Salandy, a 'mas' woman', means making essential connections to one's community. In a discussion on the aesthetic diversity of Trinidad's carnival, John Nunley points out a paradox between the individual and the community: 'That aesthetic [of diversity] is based on the principle of individuality in community – the resolution of contrasts in community and, more remotely, the cultural clashes of the past. This aesthetic fusion liberates *mas*, deploying its energy through the individual and the social body simultaneously. Herein lies the magic of Carnival, its mystery' (Nunley, 1988, p. 116).

Clary Salandy operates at grassroots level, in her community, creating alliances across local and global boundaries, practising what Stuart Hall has called a 'globalization from below' under conditions in which she and her carnival colleagues have a significant degree of agency (in Khan, 2005, p. 71). She and her collective mas camp Mahogany are slowly but surely transforming the place of carnival in London by performing an 'aesthetic fusion' which is paving the way for future generations while working locally in the present.

Notes

1. Minshall's work, which includes the opening ceremonies the Atlanta and Barcelona Olympics, is widely recognized in numerous publications as is Wayne Berkeley's, whose archives were recently purchased by the Heritage Library, in the National Library of Trinidad and Tobago, a significant acknowledgement of its importance that assures its legacy within his own country. See Ferris (2009) for a discussion on Berkeley's digital archive.
2. 'Midnight Robbers: The Artists of Notting Hill Carnival' opened at City Hall in London in September 2007 and ran through October. The other artists are: Lawrence Noel, a carnival elder who brought the first costume band on the road in 1973, Carl Gabriel, a carnival artist who actively runs carnival workshops and is acknowledged as the wire-bending artist of Britain (wire-bending is a method of costume making from Trinidad), and Ali Pretty, who is the artistic leader of Kinetika Arts International. Pretty continues to design costume bands but they have not appeared at Notting Hill since 2005. For more details on these artists see Ferris and Tompsett (2007), *Midnight Robbers: The Artists of Notting Hill Carnival.*
3. This information was retrieved on 7 April 2006 from the Poison TT website, run by www.duchessevents.com. Poison TT appears to be defunct, based on a variety of web-based information retrieved on 3 May 2009. Poison UK, however, continues to be active and thriving.

References

Aching, G. (2002). *Masking and power: Carnival and popular culture in Trinidad.* Minneapolis: University of Minnesota Press.

Alibhai-Brown, Y. (2001). *Imagining the new Britain.* New York: Routledge.

Boyce Davies, C. (2007). *Left of Karl Marx: The political life of black communist Claudia Jones.* Durham, NC: Duke University Press.

Cohen, A. (1993). *Masquerade politics: Explorations in the structure of urban cultural movements.* Berkeley: University of California Press.

Crozier, C. (2005, July). *Processes of carnival art.* Paper presented at the Arts and Cultural Politics of Carnival Research Seminar, Oberman Center for Advanced Studies, University of Iowa.

Ferris, L. (2009). The challenges of archiving and researching carnival art. *Theatre Survey, 50*(1, May), 127–134.

Ferris, L., & Tompsett, A.R. (2007). *Midnight robbers: The artists of Notting Hill Carnival.* London: Carnival Exhibition Group.

Funk, R. (2003, September 16). Errol Hill 1921–2003. *Kaiso Newsletter No. 42*. Retrieved May 15, 2006, from http://www.mustrad.org.uk.

Hall, S. (1999). Thinking the diaspora: Home-thoughts from abroad. *Small Axe: A Journal of Criticism*, *6*(September), 1–18.

Hill, E. (1997). *The Trinidad carnival: Mandate for a national theatre.* London: New Beacon Books.

Khan, N. (2005). Building from below. In R. Fenton & L. Neal (Eds.), *The turning world: Stories from the London International Festival of Theatre* (pp. 61–71). London: Calouste Gulbenkian Foundation.

Nunley, J. (1988). Masquerade mix-up in Trinidad carnival: Live once, die forever. In J. Nunley & J. Bettelheim (Eds.), *Caribbean festival arts* (pp. 84–117). Seattle: University of Washington Press.

Nurse, K. (2004). Globalization in reverse: Diaspora and the export of Trinidad carnival. In M.C. Riggio (Ed.), *Carnival! culture in action – The Trinidad experience* (pp. 245–254). New York: Routledge.

Owusu, K. (1986). *The struggle for black arts in Britain.* London: Comedia Publishing Group.

Poison UK. (2006). Retrieved on April 7, 2006 from www.poisonuk.com. [Note: this website is active (3 May 2009) at www.poisonuk.com/joomla/]

Riggio, M.C. (2004). Time out or time in? The urban dialectic of carnival. In M.C. Riggio (Ed.), *Carnival! culture in action – The Trinidad experience* (pp. 13–30). New York: Routledge.

Salandy, C. (2005). A stretch of the imagination. Artist statement exhibited on studio wall, Mahogany Arts, Harlesden, London.

Salandy, C. (2006, January). *The art of carnival.* Public lecture presented at the Roy Bowen Theatre, Ohio State University, Columbus, Ohio.

Salandy, C. (2007). Clary Salandy: The interview. In L. Ferris & A.R. Tompsett (Eds.), *Midnight robbers: The artists of Notting Hill Carnival* (pp. 28–33). London: Carnival Exhibition Group.

Sherwood, M. (1999). *Claudia Jones: A life in exile.* London: Lawrence and Wishart.

Tompsett, A.R. (2002). Wings soar, sequins shimmer: Textile arts in the Notting Hill Carnival. *Embroidery, 53*(September), 39–41.

Tompsett, A.R. (2005). 'London is the place for me': Performance and identity in the Notting Hill Carnival. *Theatre History Studies, 25*, 43–60.

Tompsett, A.R. (2006, January). *Perspectives on London's Notting Hill Carnival: A grass roots art form with a global profile.* Public lecture presented at Roy Bowen Theatre, Ohio State University, Columbus, Ohio.

Carnivals against capital: radical clowning and the global justice movement

L.M. Bogad

Department of Theatre and Dance, University of California, Davis, USA

This essay looks at an international performance phenomenon, which I refer to as *tactical carnival*, that has developed as a tactic in the toolbox of the burgeoning global justice movement. This movement has been more accurately described as a 'movement of movements' due to its great diversity in geography, identity, and ideology ('One no, many yeses' is one of its main slogans). As connections and coalitions are forged between Bolivian miners, American anti-corporate activists, Polish organic farmers, etc., organizers have begun to coordinate a celebratory form of protest that involves *unpermitted* street parties/processions that occupy public space, both to assert movement identity and importance and often to disrupt state or corporate events/daily business. Movement organizers and writers use the term 'carnival' to label these explicitly oppositional events, at which flamboyant costumes, dance, puppets, tricksterism, samba bands and other musical groupings can all be seen. They also seem to refer to ideas about 'carnival' that may, to some scholars, seem romantic or overly idealist: nevertheless, these activists are attempting to deploy the ideal of carnival in a practical, experimental way on the street, to create a new, twenty-first century kind of 'carnival' that is not calendrically nor spatially circumscribed or permitted by the state but declared and embodied by a movement that identifies itself as global, anti-corporate and anti-authoritarian.

> Run Away From the Circus – Join the CIRCA! (Clandestine Insurgent Rebel Clown Army (CIRCA) recruiting slogan)

> CIRCA is the only army in the world in which General Strike outranks Private Property. (Colonel Oftruth)

Edinburgh, Scotland, during the Carnival for Full Enjoyment, 4 July 2005

Police in black riot gear stand shoulder to shoulder. In the manner of a classical phalanx, their clear, body-length plastic shields form an impervious wall across the width of the city street. Those shields have just been used to shove people down the road and assert police control of the space. Their helmets, visors down, provide protection and make their faces harder to see. Their boots and fireproofed full-body armor are imposing. They are ready to preserve public order.

The police are confronted by a disorderly gaggle of men and women in chaotic facepaint, dressed in a tensive mixture of second hand military gear and clownish, fuzzy, garish pink and green frills. These are the tricksters of the Clandestine Insurgent Rebel Clown Army, or CIRCA. Far from seeming intimidated by the powerful policemen who have just shoved them and are now blocking the way, the CIRCA folk seem overjoyed to see them and hail them as friends to play games with. The clowns begin scrubbing the policemen's boots with their feather dusters, then breathing on their shields to fog them up and scrubbing the shields as well.

One of the main rules of improvisation theatre is the idea of 'yes, and'; one never negates a performing partner's idea or proposition, but rather agrees and adds a new creative thought to the mix. The rebel clowns jump about, following each other in an agreeable 'yes, and' ethos of improvisational frolicking in front of the police line. They simply refuse to acknowledge the very clear and stark NO of the disciplined, uniformed, and armed police line, asserting their own ridiculous and enthusiastic YES AND over and over with each moment. Finally one intrepid trickster, aptly named Trixie, kisses one of the police shields (see Figure 1). Her kiss is so enthusiastic and vigorous that she smears her clown makeup and lipstick all over the shield. She then goes from shield to shield, all the way up the phalanx-line, kissing and leaving her smeary mark on each one as she goes. The police stay in formation, some disturbed, some impassive, and some amused and surprised by this

Figure 1. CIRCA Clownbattants confront the riot police, Edinburgh, 4 July 2005. Photo by Matthew Dutton.

paradigm-shifting kiss. One says 'Step away, from the shield please', while another can clearly be seen to be smiling.

Trixie needs to re-apply her lipstick in order to continue her loving assault, and while doing this she is asked by Zoe, a CIRCA videographer, why she is doing what she's doing. In a high pitched voice she earnestly says, 'Because I love them! I love the police! They're our friends'. She then runs off to kiss more shields.

Trixie then begins drawing smiley-faces with her lipstick and saying 'Yey!!' Soon she is writing 'Yey!!' on some of the riot shields, while the other clowns cheer and clap. The images of her kisses and incongruous lipstick happy faces are broadcast and printed all over the world the next day. There is a marked gender element to this interaction – the male clowns aren't kissing any shields, but they are scrubbing away solicitously, saying things like 'Oh, it's a mess, it's very bad – don't worry! I'll clean it, you won't get in any trouble!'

Soon the rebel clowns form a line facing the police, and begin doing a motion/ sound all together: bending at the hip, arms and hands extended, and going 'Shhhhh', then standing back up while making a 'Whooo' sound. Their relentless group improv had led them to an ambivalent, absurd gesture; the clowns were simultaneously kowtowing to the police and shooing them away. After five or six repetitions of this movement, the clowns began jumping up and down, yelling 'Yippie!' At that precise moment, the police line broke. First one, then two, then the entire squad pulled away, formed two files, and began jogging down the road at double time. The clowns ran along with them briefly, and a great cheer rose from the mass of non-clown protesters and bystanders, who followed the police for a few yards and then began to celebrate in the street.

It is probably prudent at this point to note that, in fact, the police line may not have been broken by clown magic. One can imagine the rear-rank conversation between the officers of the unit, or by walkie talkie to headquarters: '*Look, we've become sucked into some kind of performance art here, and we'd really like to go elsewhere so we can fight crime . . . well, look to be fair, our training doesn't include improv theatre . . . requesting permission to withdraw*'. However, the constant improvisation of the clowns made such a poetic moment, where it appeared that clowns could dispel the intimidating power of the state through fearless silliness and serious play, possible. As one clown said, 'They loved us so much, we asked them to go away, and they did'. This was not 'street theatre', per se, but a form of improvisatory and creative direct action. ('Direct action' is a term for activist methods that directly confront and, perhaps, disrupt or change sociopolitical processes, as opposed to indirect activism that seeks to lobby for change through influencing politicians or swaying public opinion.) The clowns immediately began to fill the authoritarian gap created by the policemen's withdrawal – power abhors a vacuum, after all – by directing traffic, including police wagons, with their feather dusters and giving passing vehicles the CIRCA Clown Salute (which, according to CIRCA regulations, is thumbing one's nose with a big smile on one's face). For the most part, however, they continued playing games as they had been before the police arrived to demarcate the space.

This absurd face-off between two groups of masked, uniformed performers occurred during the 'Carnival for Full Enjoyment' of 4 July 2005. This carnival was not a state-sanctioned cultural steam-valve, occurring within universally recognized temporal bounds based on the agrarian calendar. Rather, it was declared by members of the global justice movement as a festive and defiant event. While it was in part a response to the G8 summit happening a few miles away at Gleneagles, it was not purely negative and reactive in character, but an attempt to open up a space for anti-authoritarian, egalitarian and participatory celebration without state permits or sanction. With slogans such as 'No Wage Slavery', 'No Benefits Slavery', 'No Army Slavery', and 'No Debt Slavery', the Carnival for Full Enjoyment called, in a press release, for workers to 'Phone in Sick and Join the Carnival':

> Flex, temp, full-time, part-time, casual and contortionist workers, migrants, students, benefit claimers, New Dealers, work refusers, pensioners, dreamers, duckers & divers... Bring drums, music, banners, imagination for action against the G8 that expresses our resistance in work, out of work and wherever we live. Assert our desires for FULL ENJOYMENT with fun in the city – and begin to make capitalism & wage slavery history... *On 4 July we can take action and experience – if only temporarily – what life could be like if we got the bosses off our backs.*
>
> The Carnival for Full Enjoyment involves both local people and people from round Britain – and beyond. We are making it a carnival because life should be more fun for those of us who labour in underpaid insecure jobs, in casual and agency work, or on 'New Deal' schemes. There should be more joy for the unemployed trying to survive on a few quid a week and for those of us juggling childcare and debt... *Get together with friends and set your sights. Bring what you'd want to find, and most of all bring imagination and passion. Diversity and creativity is our strength.*
>
> We know many other groups locally are glad the Carnival is happening. For example Esther MacDonald of Edinburgh Claimants told us: 'Every week we give advice and solidarity to people resisting benefits cuts and harassment and to people standing up to debt collectors. We welcome the Carnival for Full Enjoyment. From Niddrie to Nigeria, poor people need to organise ourselves to take the world's resources into the hands of the world's people'.
>
> Our purpose is to oppose those who devastate our communities through economic exploitation. We will make clear our resistance to organisations which thrive on the poverty and debt in which many of us find ourselves. Because our society is mainly based on using people to make money for others, there are many such organisations to choose from.
>
> *We are in favour of direct action because marching around with placards can be safely ignored by those who control exploitation.* We advocate direct action against the institutions which exploit the majority... We invite workers from Standard Life and all other corporations to join the Carnival. Take an extended lunch-break, phone in sick! Join us in opposing casualisation, the intensification of work, attacks on pensions and conditions. Edinburgh call centre worker David Gillies said 'At my work they monitor our calls and threaten us with disciplinary action if we spend too long speaking to the caller. The company only cares about profit. I'm taking the day off on the 4th to join the Carnival'.
>
> *The carnival is a celebration of how good life can be, and at the same time a statement against those who spoil it for the majority.* (Carnival for Full Enjoyment, my emphasis)

The organizers of the Carnival explicitly presented it as an anti-capitalist, anti-authoritarian event with links between local workers and activists and a larger global movement. The Carnival was to be both a rejection of state authority (very significantly, no permit was requested for this event), an affirmation of the joy of solidarity and resistance (particularly resistance to the regulation of everyday behavior in increasingly privatized and controlled public space), and an alternative to what was seen as the more staid tactic of 'marching around with placards'.

It was also seen, both by some protesters and the authorities, as a possible opportunity for direct action to disrupt the proceedings of corporations and 'business as usual' in the city. A massive police force was concentrated in Edinburgh to protect banks and other corporate offices from vandalism or nonviolent disruption such as sit-ins, and to prevent the reveling, disorderly crowds from blocking commercial traffic on the streets. The government brought in 10,000 police from all over the UK to Edinburgh in order to respond to this threat to security and public order. Police from London, Manchester, and many other cities filled the streets, blocking intersections, surrounding, searching, arresting, chasing and breaking up large groups of protestors and keeping them separate. There was a great deal of game playing, samba bands, festive costumes, dancing, and revelry in the street, and even a huge anarchist black cat-puppet; however, there was also plenty of fleeing, regrouping, cat-and-mouse and, on one street, even violent confrontation between protestors and police.

This was not exactly a carnival in the Bakhtinian sense, where the laughter is ideologically ambivalent, where all participate, and where the event is demarcated and approved of by the authorities even though there is always the risk that the licensed fools may go too far. Nevertheless, the movements that organize such events are to some degree influenced by the ideas and writings of Bakhtin and Rabelais (especially in Western countries), and they have theorized such oppositional carnivals as an important tactic in the struggle against corporate globalization.

Bakhtin and the carnivalesque

Mikhail Bakhtin (1895–1975), a Russian literary scholar and philosopher who struggled in obscurity under the reign of Stalin, and whose major works were only published in the last fifteen years of his life, celebrated the carnivalesque, mass-participatory freedom and anti-authoritarian laughter that he found in the writings of Rabelais:

> Carnival does not know footlights, in the sense that it does not acknowledge any distinction between actors and spectators. Footlights would destroy a carnival, as the absence of footlights would destroy a theatrical performance. Carnival is not a spectacle seen by the people; they live in it, and everyone participates because its very idea embraces all the people. While carnival lasts, there is no life outside it. During carnival time life is subject only to its laws, that is, the laws of its own freedom. (Bakhtin, 1968, p. 7)

The 'carnivalesque', and its liberating qualities for the 'lower orders' of society, includes abuse and laughter, which degrade at the same time that they renew. Grotesque realism exaggerates the material body and the 'lower bodily stratum', inverting the hierarchies of elite taste and decorum and the symbols of hierarchy. This is a frenetic, celebratory and ideologically ambivalent performance mode which breaks down the bodily boundaries of the idealized bourgeois individual, 'polluting' and collectivizing the human condition in a joyous, outrageously humorous demonstration that has some potential for rebellion (Bakhtin, 1968).

'Carnivals against capital'

This essay looks at an international performance phenomenon, which I refer to as *tactical carnival*, that has developed as a tactic in the toolbox of the burgeoning global justice movement. This movement has been more accurately described as a 'movement of movements' due to its great diversity in geography, identity, and ideology ('One no, many yeses' is one of its main slogans). As connections and coalitions are forged between Bolivian miners, American anti-corporate activists, Polish organic farmers, etc., organizers have begun to coordinate a celebratory form of protest that involves *unpermitted* street parties/processions that occupy public space, both to assert movement identity and importance and often to disrupt state or corporate events/daily business. Movement organizers and writers use the term 'carnival' to label these explicitly oppositional events, at which flamboyant costumes, dance, puppets, tricksterism, samba bands and other musical groupings can all be seen. They also seem to refer to ideas about 'carnival' that may, to some scholars, seem romantic or overly idealist: nevertheless, what is fascinating is that they are attempting to deploy the ideal of carnival in a practical, experimental way on the street, to create a new, twenty-first century kind of 'carnival' that is neither calendrically nor spatially circumscribed or permitted by the state but declared and embodied by a movement that identifies itself as global, anti-corporate and anti-authoritarian.

I posit that the goals of tactical carnival are:

- to declare and occupy a joyous, participatory and semi-anonymous, relatively safe place for power inversions/subversions. 'Celebrity' has been explicitly denounced in some movement literature, in favor of the relative anonymity of the mass. These spaces are also meant to be non-dogmatic/sectarian, a more open place for wider participation. The hope is that more people will join the movement when a space for this kind of joyful participation is opened up.
- to put a friendly face on the movement as a way to interrupt what I refer to as the *hegemonologue* of the corporate media and state rhetoric, which often demonizes and others activists as crazed, nihilistic hooligans (as quoted in Shepard, 2005; Bogad, 2006). The idea is to insert images that at least partially disrupt or disharmonize the barrage of negative images (for example, a clown kissing a riot shield juxtaposed with the usual images of street melee and property damage), and to replace the usual 'story of the battle' (street fights, vandalism, etc), with the 'battle of the story' in which colorful and creative costumes, dance, music, performance and improvised interactions give a new look at the movement and its agenda (D. Solnit, personal communication, 2 July 2005).
- These events also attempt to interrupt another aspect of the *hegemonologue*, which is that of the rhetoric of the inevitability of corporate globalization, by demonstrating that better alternatives are possible.
- to key an experimental mode in which new ways to play with and around power can be tested. The idea is to develop less obvious and predictable ways to interact on the street with agents of the state, corporations and passersby. Much of the creativity is intended to have the effect of dispelling fear and tension during confrontations with massive police presence, for example.

- to create an celebratory culture of active defiance, as an alternative to the everyday life experience of many people – in response to a widespread frustration that many participants feel – regarding their official relegation to the role of consumers of culture and spectacle rather than creators/spect-actors (Boal, 1985).

I examine this global phenomenon through a local particular: the G8 protest in July of 2005, with a further focus on the Carnival for Full Enjoyment in Edinburgh on 4 July as participated in by the Clandestine Insurgent Rebel Clown Army (CIRCA). I will draw on press clippings, videos, interviews with CIRCA members, movement literature and my own experience as a participant/observer to examine how CIRCA theorized and actualized its own participation in the tactical carnival around the G8 in July 2005.

'Carnival' is clearly a referent for this movement and many of its actions, even if Bakhtin would not recognize these events as Rabelasian. On 18 June 1999, the opening day of the G8 summit in Koln, Germany, a 'Carnival Against Capital' was declared by the global justice movement as an 'international day of action, protest and carnival aimed at the heart of the global economy' (Ainger, 2004, p. 33). This proposal identified 'capitalism as the "root of our social and ecological problems", and [was] taken up by the People's Global Action network, translated into seven languages, and distributed by email and post to thousands of groups worldwide' (Notes From Nowhere [NWN], 2003, p. 184). On that day, in the financial district of London, 10,000 people gathered in the street, playing 'Volleyball with inflatable globes and danc[ing] to samba rhythms in the spray of a waterspout from a damaged fire hydrant'. In order to avoid being penned in by the police, to keep movement fluid, and to add a colorful flair to the event, organizers gave participants color coded masks (red, yellow, blue, etc) to wear; a note inside the masks suggested that the wearers follow the flags that matched their masks when the time came. Sure enough, at one moment, colorful flags went up and soon streams of masked revelers were running, following the flagbearers on zigzagging routes through the narrow streets of the City of London. Of course, this event had not been permitted by the state, and the anti-capitalist and anti-authoritarian ideology of the gathering asserted itself through direct action as the day progressed:

> ... by the end of the day a group of the protesters had invaded and trashed the ground floor of the London International Financial Futures Exchange, three McDonalds had their windows broken, two people had been run over by police vans, and riot police were charging in. The sight of anarchy hitting the world's largest financial center prompted newspaper headlines that denounced the protesters as 'evil savages', an ignorant 'unwashed horde' hell-bent on turning a 'carnival into a riot'... the carnival goers in London – the majority of whom had been nonviolent in actions and intent – were members of a far larger, invisible but international constituency organizing around a common enemy: globalization. (Ainger, 2004, p. 33)

This Carnival Against Capital was hardly limited to London. Simultaneous protests of similar nature took place 'against global capitalism, the international financial system, and corporate power... in 43 countries around the world' (Ainger, 2004, p. 33). For example, in Nigeria, a 'Carnival of the Oppressed' bought 'nearly 10,000 Ogoni, Ijaw, and other tribes together in closing down the country's oil capital, Port Harcourt... meanwhile in Koln, the Intercontinental Caravan, made up of 400 Indian farmers and other activists from the global South, plan[ned] to conclude its tour with a Laugh Parade but police detain[ed] 250 of them before they [got] the chance to guffaw at the G8' (NFN, 2003, pp. 185–177). In fact, this global event was so disturbing that the FBI listed 'Carnival Against Capital' as a terrorist group to be watched in their memo of 11 May 2001, four months before the Al Qaeda attacks of 11 September of that year (FBI, 2001). This is ironic in that 'Carnival Against Capital' was not an organization but rather a concept, a call to action, an invocation of the idea of the subversive and celebratory '*carnivalesque*' that resonated around the world for the global justice movement.

There are many other examples of the movement's invocation of the idea of 'carnival' for its actions. The anti-World Trade Organization protests that shut down the city of Seattle in 1999 were also conceived, in part, as carnivalesque, with colorful costumes, giant puppets, dancers and marching bands, and people dressed as sea turtles and butterflies, playing and interacting with more conventional union members and other activists. The protest against the Free Trade Area of the Americas (FTAA) conference in Quebec in 2001 was also dubbed a 'Carnival Against Capital'. There is a neckerchief, sewn by Argentinian women textile workers who took over their own factory during the financial/economic collapse in that country, that is worn by many movement activists, including members of CIRCA. The neckerchief, produced in red, yellow, pink, and orange colors, has a smiling face printed on one side: one can wear it so that the mass-produced smile covers the lower half of one's face. On the other side a pattern of a chain-link fence covers an unhappy face, so the wearer has the choice of looking free and happy or fenced-in and sad. One can imagine a mass of people, filling up a public space, dancing or playing games and making music, while wearing these masks. The effect might be ideologically unclear, yet eerie. On the edges of the neckerchief, in English, Spanish, German and French, is a passage that explicitly states what 'Carnival' means for the global justice movement:

> We will remain faceless because we refuse the spectacle of celebrity, because we are everyone, because the carnival beckons, because the world is upside down, because we are everywhere. By wearing masks we show that who we are is not as important as what we want, and what we want is everything for everyone. ('We will...')

This passage shows some of the key ideas that the global justice movement embraces from their creative interpretation of the carnivalesque. The idea of anonymity is advantageous for concealing identity from the authorities, but also as a way to celebrate the undifferentiated mass and combat the worship of individual celebrity/cult of personality by the mass media and pop culture. This reflects the Bakhtinian distinction between the carnivalesque mass of undifferentiated bodies, versus the discrete, separate, and closed-off bourgeois individual body uplifted by mercantilism and capitalism. Anonymity also serves as a tool for tactical players; the protesters lack the strategic power of the state or the corporations, but anonymity enables them to negotiate the urban landscape, maneuver, play, and perhaps escape while in full view of the dominating authorities (de Certeau, 1984). The demand for 'everything for everyone' echoes the millenarian, utopian poetics of the Rabelasian carnival, also expressed in the 'hidden transcripts' of the oppressed of many peasant cultures (Scott, 1985, 1990) and the rhetoric of the Diggers (Hill, 2002). 'The carnival beckons' this movement to action of a celebratory kind.

Many groups have formed with the goal of contributing art, festive costumes, and music to these carnivals of resistance. One group explains their savvy theorization of themselves right in their name: Tactical Frivolity. Musical bands are a crucial aspect of these gatherings; one such group playfully named itself Reclaim the Beats, after the seminal movement group, Reclaim the Streets. The choice of music varies, but there is a great deal of influence from the carnivals of Brazilian and the Caribbean communities, showing that tactical carnival is not only motivated by a reference to medieval European traditions. While the (now-defunct) Seattle-based Infernal Noise Brigade (www.infernalnoise.org) played a postmodern blend including 'elements of drumline, taiko, Mughal and North African rhythms, elements of Balkan fanfares, breakbeats, and just about anything else' (Infernal Noise Brigade) and dressed in an almost paramilitary uniform, the UK-based Rhythms of Resistance say about themselves:

> Whilst people often refer to us as a 'Samba Band' we actually have more affinity with the Afro Bloc parading drum bands that emerged in the mid 70s in Salvadore, Bahia in Brazil. Bands such as Ile Aye and Olodum formed as a political expression of black awareness, resisting economic exclusion. Coming out of some of the poorest urban communities, Afro blocs became a mobilising focus on picket lines and marches.
>
> As they put it, they played as 'a force of resistance and source of self confidence'. The growth of Schools of Samba both in Brazil and all over the world since the 80s, is largely a result of the commercialisation of this culture of resistance.
>
> Rhythms of Resistance formed as part of the UK Earth First action against the IMF/World Bank in Prague in September 2000. A Pink and Silver carnival bloc, focused around a 55 piece band, detached itself from a march of 67,000 and outmanouvered police resources defending the IMF annual summit. With an international 'black bloc' and a large contingent from the Italian movement, 'Ya Basta', three diverse forms of direct action worked towards a common goal and resulted in the shut down of the IMF summit.
>
> Building on the success of S26 [the Prague protest], more bands are forming, playing the same rhythms and with the same approach. The Electric Blue band formed in Amsterdam, for the Rising Tide Actions against the Cop6 Climate Conference in November 2000 and joined together with 10 drummers from Rhythms of Resistance to form a 65 piece band.
>
> With bands forming in across Europe and beyond (at least 2 in the US), an international network of percussive resistance to the march of capitalism is now emerging.
>
> *Street carnival is the vital component of protest and life and fun – use your imaginations, connect and network, build instruments and costumes, learn our tunes and distribute them noisily through the world!!* (Rhythms of Resistance, my emphasis)

Rhythms of Resistance exemplifies the anti-capitalist movement's awareness of and perhaps problematic appropriation/exchange of oppositional culture and tactics across borders of nationality, race, class, and privilege. Just as power and commerce are circulating around the world at an accelerated rate in the era of globalization, so are traditions and innovations in cultures of resistance. The same internet technology that enabled the global movement to coordinate the Carnival Against Capital in 43 different countries around the world also enables such an exchange.

From the text on their website, it is clear that Rhythms of Resistance, which has inspired many such groups for protests in other countries, has embraced the idea of 'carnival' as a positive and uplifting mode of defiance and advocacy. They also perform their music at protests in order to foster group 'confidence' and defiant joy, and simply to help make the protest a good time for all who join in.

Why is 'carnival' such an important referent for this 'movement of movements'? Carnival, as conceived by Bakhtin, does have an edge. It suggests a possibility of riot or rebellion, that the licensed foolery might get out of hand and turn into outright revolt; this may appeal to some movement groups. However, more importantly, some modern anarchists are particularly drawn to the idea of carnival because it appeals to their egalitarian ideology and participatory, do-it-yourself (DIY) ethos. Bakhtin's all-too-familiar statement that 'carnival knows no footlights' is eagerly echoed in the text on the above-quoted neckerchief, and in the key movement text *We Are Everywhere: The Irresistible Rise of Global Anticapitalism*. In that book, in the 'Carnival' section, under the subheading of 'Participate, Don't Spectate', the famous Bakhtin passage about footlights is quoted in full, followed by:

> Passivity disappears when carnival comes to town, with its unyielding demand for participation... It is a moment when we can break free from the alienation that capitalism enforces in so many ways... Carnival denies the existence of experts, or rather, insists that everyone is one... it demands interaction and flexibility, face-to-face contact and collective decision-making, so that a dynamic and direct democracy develops – a democracy which takes place on the stage of spontaneously unfolding life, not raised above the audience but at ground level, where everyone can be involved. There are no leaders, no spectators, no sidelines, only an entanglement of many players who do their own thing while feeling part of a greater whole. (NFN, 2003, pp. 177–178)

This perhaps idealistic desire that, in carnival, the mass is involved actively and that no one is relegated to the role of passive spectator/consumer, particularly appeals to a movement that views the mainstream news and entertainment media with great skepticism, that is influenced by Guy Debord's concept of the deadening, pacifying, and self-perpetuating 'society of the spectacle' (Debord, 1995), and that advocates and practices a participatory, 'do-it-yourself' form of political direct action and communication (for example, though zines and Indymedia). Many members of the movement, while not explicitly anarchist, do embrace the term coined by the movement in Argentina: 'horizontalist'. This egalitarian idea of 'horizontality' calls for a minimization of the concentration of power in any single person or group's hands, and Bakhtin's idea of the carnival, which has been critiqued by scholars, nevertheless appeals strongly to such ideological desires and agendas. To distinguish tactical carnivals from the carnivals that occur in Rio de Janeiro or Notting Hill, members of the anti-capitalist Notes From Nowhere collective wrote:

> What carnivals remain in most parts of the world have themselves become spectacles – specialist performances watched by spectators – with police lines and barriers placed between the parade and audience. Thus the vortexed, whirling, uncontrollable state of creative chaos is shoe-horned into neat straight lines and rectangles. A visit to many contemporary carnivals sanctioned by the state (such as Carnaval in Rio de Janeiro, or the Notting Hill Carnival in London), where consumption and corporate sponsorship has taken over from the creativity and spontaneity is enough to illustrate how carnival under capitalism has lost its vitality. But carnival has been with us since time

> immemorial and it has always refused to die. Reappearing in different guises across the ages it returns again and again. Freed from the clutches of entertainment, the anticapitalist movements have thrown it back into the streets, where it is liberated from commerce for everyone to enjoy once again. (NFN, 2003, p. 177)

This passage reveals how important the concept of carnival is to some social movement activists, and how it has motivated the way they organize and perform in public space in critical contrast with (the activists' perception of) events such as Rio's Carnaval.

However, this form of carnival is also different from the form described and idealized by Bakhtin. It is not a yearly event that fits into an agrarian calendar and a Christian/feudal worldview and system of power; it happens when and where the movement calls for it, though often in reaction to the corporate/state events such as the meetings of the Group of 8 (G8) or the World Trade Organization (WTO). Far from being a day of licensed foolery and excess, it is often unpermitted by the state, triggering police repression. Tactical carnival comes from a complex, global coalition of social movements, and therefore defies ideological generalization. However, its laughter, while ambivalent and self-mocking at times, is not as ambivalent as that described by Bakhtin. It is joyful, but it is also satirical laughter, often coupled with a political worldview or critique, and therefore perhaps not quite as universal as Bakhtin's. It espouses not just the 'World-Turned-Upside Down' concept often described by scholars of medieval European carnival, but rather the very different slogan 'Another World Is Possible'. The difference between these two phrases is the difference between a temporary inversion of power relations and an assertion of the possibility of, and advocacy for the struggle towards, lasting and substantial progressive social change. This slogan, 'Another World Is Possible', has been used in several languages on movement proclamations and banners around the world.

This form of protest, invoking the carnivalesque, has not only been used to oppose capitalism. As Padraic Kenney discusses in his book *A Carnival of Revolution*, the grassroots, independent groups in Eastern Europe who undermined authoritarian Communist power using absurdist, carnivalesque mass protests were a key, undersung element in the collapse of Soviet power in several of the Warsaw Pact countries (Kenney, 2002).

Part of this tactical carnival model is a response to more conventional and institutionalized models of social movement protest. The goal here is to *open up* public space, with do-it-yourself group and individual creativity, rather than to merely *occupy* it with uniform marching and chanting while holding mass-produced signs. This opening of public space for a freer and more festive kind of protest and direct action is one of the main goals of CIRCA.

CIRCA: The Clandestine Insurgent Rebel Clown Army

CIRCA was formed in London in the fall of 2003. It declared itself to be an army of 'rebel clowns' who had mustered their forces to storm the Palace during President George W. Bush's state visit to the UK. CIRCA had been delighted that, after hundreds of years of having no fools or jesters in the court, that the Queen was finally opening her doors for a fool to come . . . but CIRCA was soon dismayed to hear that it was the wrong kind of fool whom Her Majesty had invited. CIRCA was seen all over the city, firing pink pretzels out of their clown cannon at the Esso building, marching

Figure 2. Recruiting card. Courtesy of the Laboratory of Insurrectionary Imagination.

behind Beefeaters, 'sneaking' though the park in their bright pink and lime green fuzzy outfits while holding single leaves in front of them for camouflage (and holding the leaves above their heads in response to police helicopters flying over). Since that time, CIRCA has been seen occupying the city of Leeds, marching on the 'Greet the Buyers' meeting in London between corporations and the US-installed interim regime in Iraq, attempting to attend the 'Republican National Clown Convention' in New York as members of the Big Top Delegation, and elsewhere. CIRCA uses a horizontalist organizing model as it practises a form of nonviolent direct action that joins collective buffoonery with satiric performance.

CIRCA can often be seen in the street moving as a mass body – in tight-knit gaggles of ten or twelve, they may march in mock-disciplined but surprisingly tight pseudomilitary formation, then suddenly break off into total clownarchy – antics and improvisation with passersby. Suddenly, and with no cue, the rebel clowns begin moving like a tightly clustered school of fish, making the same sound and gesture, then changing direction simultaneously and making an entirely different sound and gesture (this is called *fishing*). The unpredictability of CIRCA's behavior, and the constant collective shape-shifting, come from hours of group practice and training. In their huddled/cuddled group movements, CIRCA evokes the Bakhtinian idea of the carnivalesque mass body as opposed to the idealized, individualistic, and discrete/discreet bourgeois body. CIRCA seems to collectively embody the very-Bakhtinian concepts of Notes From Nowhere:

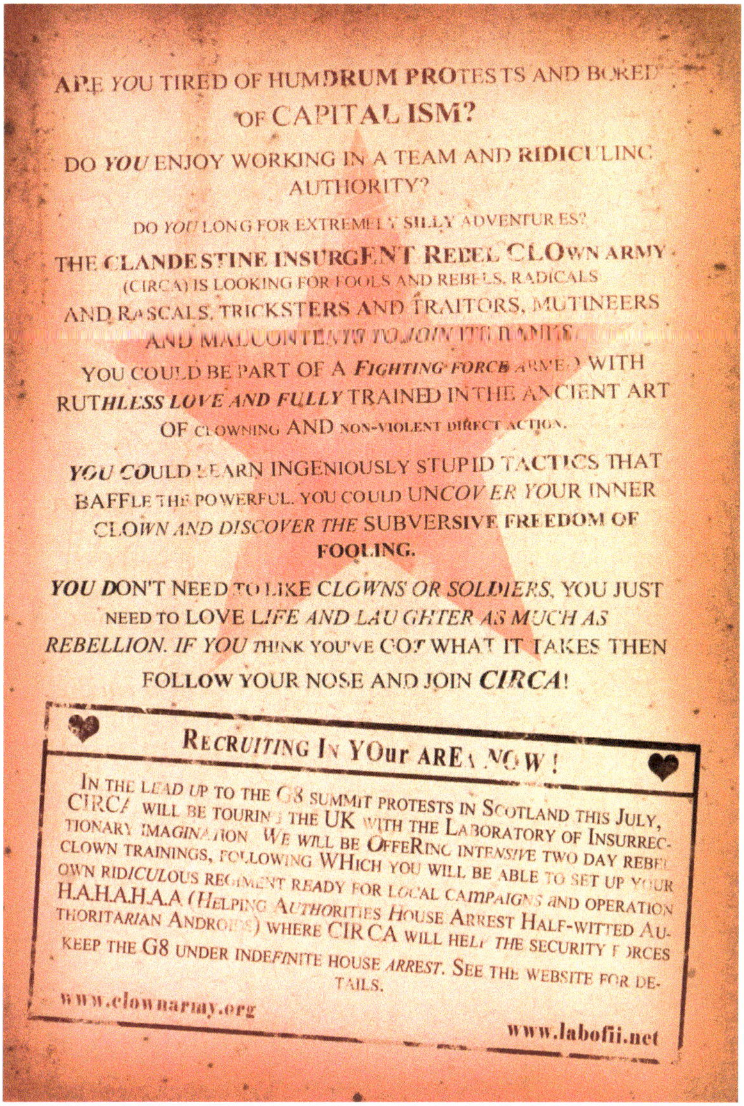

Figure 3. Recruiting card. Courtesy of the Laboratory of Insurrectionary Imagination.

The pleasures of the body have been banished from the public sphere of politics and the excitement of the erotic pushed into the narrow private confines of the sexual realm. But carnival brings the body back to public space, not the perfect smooth bodies that promote consumption on billboards and magazines, not the manipulated plastic bodies of MTV and party political broadcasts, but the body of warm flesh, of blood and guts, organs and orifices.

During carnival the body sticks its tongue out as far as it can, it laughs uncontrollably, sweats and farts as it dances in the heat of other bodies. It's a body that refuses the static images of itself developed by capital, frozen in immortal youthfulness, aloof from natural cycles of eating and shitting, being born and decomposing. In carnival the body is always changing, constantly becoming, eternally unfinished. Inseparable from nature and fused to other bodies around it, the body remembers that it is not a detached, atomized being,

> as it allows its erotic impulse to jump from body to body, sound to sound, mask to mask, to swirl across the streets, filling every nook and cranny, every fold of flesh. During carnival the body, with its pleasures and desires, can be found everywhere, luxuriating in its freedom and inverting the everyday. (NFN, 2003, pp. 175–176)

While CIRCA members rarely excrete on the street, they do move in groups (or 'gaggles' as the clowns prefer to call them) and in ways that celebrate the mass body over the individual, and earthiness and silliness over commercial standards of beauty and respectability.

> We see innovative forms of action as key for building dynamic social movements, but realize that the psyche is as important a site for struggle as the street. CIRCA believes that a self-destructive tendency within many social movements is forgetting the inner work of personal liberation and transformation. This is an area our rebel clown trainings work on deeply, while also providing creative tools to confuse and befuddle authority. (CIRCA, 2005a)

As CIRCA members Isabelle Fremeaux and Hilary Ramsden describe in their essay 'We Disobey to Love: Rebel Clowning for Social Change', CIRCA training sessions aim to guide participants towards discovering and developing their own clown persona, learning to work together in groups without leaders, and joining freely and fearlessly in improvisation under pressure (Fremeaux & Ramsden, 2007). Finding one's clown involves coming up with a name, a sensibility, a physicality and a costume and makeup style. But it is also a much longer and deeper process that involves a great deal of thoughtful/playful exploration. Putting on the makeup before an action is a crucial part of the transformation, the re-entry into one's alternate clown persona. This celebration of individual creativity and identity through the development of one's own clown can hopefully enable CIRCA members to express themselves in the moment and mode of carnival while still feeling part of a larger group identity.

CIRCA hopes to open up the spaces that they move in, to shift the paradigm or change the rules of behavior and engagement. They hope to bring the tactical carnival with them as they go, switching an event into what I refer to as the 'key of clown' though absurdist cues and gestures. The spirit of playfulness can often be infectious, and 'civilians' may be brought into the improvised games in the street – e.g. several dozen clowns coming across a speed bump and treating it as a nigh-unsurpassable obstacle that can only be climbed over through a great deal of team work and slapstick. Like police visors and shields, the face makeup and costumes of a mass of rebel clowns evokes a range of responses. However, it is the openness, willing vulnerability and fearlessness of the clowns that CIRCA hopes will prove infectious in the carnival spaces they create and bring with them as they navigate the city. As the clowns greet the police as 'friends' and fail to either melt away in fear or raise the tension in anger, a shift in the paradigm and pattern of confrontation ensues. The true challenge is to stay 'in clown' even when conventional power relationships assert themselves. In a different neighborhood during the Carnival for Full Enjoyment, a small gaggle of rebel clowns was dancing in the streets with a samba band when they were baton charged at the blow of a whistle and a cry of 'CHARGE!' by a rank of policemen. The clowns, in a tightly coordinated, nonverbal reference to the teachings of Sun Tzu and Von Clausewitz, scattered and fled at top speed. Colonel Oftruth experimented with staying 'in clown', cheerfully confiding to confused bystanders while running away, 'We must be very dangerous!'

CIRCA follows the egalitarian spirit of horizontalism and tactical carnival in its organizing model. While as many as 150 clowns were in Edinburgh at one time (and the sight of that many rebel clowns marching was a bizarre and mind-opening vision in itself), CIRCA is divided into 'gaggles' of roughly 10–15 clowns, analogous to the 'affinity group' model of the direct action movement. In groups of this size, members are working with people they can personally get to know and trust. They can develop models of group decision-making where every voice can be heard. When larger groups of clowns need to gather, they form a Clown Council where chosen spokesclowns (!) speak for and in constant consultation with their own gaggles. This basically follows the horizontalist model of organization, but with makeup and red noses added.

There was a varied attitude towards the concept of carnival amongst the CIRCA activists. Matthew Trevelyan, at some group meetings, read quotes directly from Rabelais' *Gargantua* that he found particularly inspiring and relevant to our carnivalesque efforts. On the other hand, Jennifer Verson refused to accept the term carnival to describe CIRCA's efforts, denouncing its connotation of an escape valve for societal dissent.

> The moment we choose love we begin to move towards freedom, to act in ways that liberate ourselves and others. That action is the testimony of love as the practice of freedom. (bell hooks from *Outlaw Culture: Resisting Representation*, as quoted on CIRCA's heart-shaped 'Radical Origami' paper, CIRCA, 2005b)

During the G8 campaign, CIRCA traveled throughout the UK on a biodiesel-fueled van, towing a caravan with the words 'LABORATORY OF INSURRECTIONARY IMAGINATION' painted on it. The Laboratory was the umbrella organization for the tour, including the anti-consumerist 'Church of the Immaculate Consumption'. When CIRCA pulled up into a new city, the group would set up its outdoor performance space between van and caravan, and perform a free show that included puppetry, performance art, and video presentations, all around the themes of civil disobedience, the G8, and rebel clowning. The clowns gave out 'Radical Origami' papers, in the shape of a heart and covered with movement-oriented quotes. In the course of the show, the audience was guided in folding the heart into a dunce cap, a megaphone, and a chip (French fry) container. During intermission, organic chips were distributed to all, and CIRCA explained to the audience that the fat from the chip fryer was recycled to power the biodiesel-fuelled van. Thus the possibility of a fossil-fuel-free world was exemplified in CIRCA's practice, and playfully incorporated into the performance. In the two days after the performance, CIRCA would then lead a training for interested locals in rebel clowning and nonviolent civil disobedience. In this way, the CIRCA caravan group served as a seeder group, starting CIRCA 'gaggles' in each city that would then continue to train and practise on their own. By the time of the G8, when a Clown Council was called for Edinburgh, a total of about 150 clowns, in a dozen gaggles from all over the UK, Ireland, and Belgium, with individuals from Italy, France, and Spain, gathered together to debate and discuss a plan of action.

CIRCA's main focus was on direct action in public space, but it also hoped to speak through the media by generating absurd quotes and images that nevertheless articulated a radical critique of the G8's policies, and of capitalism, that was not being enunciated by the celebrities of the Live Aid concert (it may be for this reason that Bob Geldof himself was quoted as saying that he didn't want a 'bunch of guys dressed as clowns' showing up and ruining his show). In order to spread this critique, CIRCA called an absurdist 'press conference' for 1 July 2005. The press release, written with the earnest absurdity of clown-logic, hoped to highlight CIRCA's concerns about the social movement being co-opted by Labour Party leaders who were reportedly planning to join in the 'Make Poverty History' March in Edinburgh. As always, CIRCA also wanted to call attention to their biodiesel-powered van and the possibility it raised for a fossil-fuel-free future (of course, as has become clear in the last few years, biodiesel recycled from food grease is a fine marginal method for replacing gasoline, but is not viable as a central or long-term solution):

> CLANDESTINE INSURGENT REBEL CLOWN ARMY (CIRCA)
>
> CIRCA ANNOUNCES ITS FIRST G8 CAMPAIGN PRESS CONFERENCE
> **FREE CHIPS** TO BE DISTRIBUTED TO THE MEDIA...
>
> To the Media and Press:
>
> Please meet us at ***12:30 pm on Friday 1st July at our lovely caravan on the left hand side of Teviot Student Union, Bristo Square*** for the announcement of our upcoming **Operation, BROWN-NOSE (to take place on July 2nd)**... [see our Communique below].
>
> At this briefing, we will explain our immensely important role in support of the Make Poverty History march. We will then demonstrate the techniques of radical clowning with which we will protect the march – and we will receive an honorary salute from the cannons at Edinburgh Castle, showing the Queen's approval for our campaign.
>
> **Free chips will be served!** The chip fat will power our biodiesel van to Gleneagles! If you eat the chips you will help our movement move forward!! Hee hee hee!!!
>
> With all due respect,
>
> CIRCA

The communiqué itself read as follows:

> COMMUNIQUE #8.86
>
> CLANDESTINE INSURGENT REBEL CLOWN ARMY (CIRCA)
> Re: Operation BROWN-NOSE
>
> To all our rebel friends in the global justice movement – our brave, beautiful, beatific, bracelet-wearing brethren bent on beating back and baffling the behemoth of Babel that is the Gathering of Eight...
>
> We are so proud of you, so in love with you... we share your compassion and desire to TRULY make poverty history. We admire your realization that the only real way to end poverty, in the global South or in the suburbs of Edinburgh, is to stop the Great Eight's amusingly antisocial habits – their rather nasty slapstick routines – kicking the poorer nations with war, arms sales, and bullying trade policies, manipulating their markets and plundering their resources while dangling the crumbs of 'debt forgiveness'. We know the G8, far from being thanked for 'forgiving' these illegitimate debts,

should **beg forgiveness** from the Global South for their ongoing crimes and depredations . . .

Beyond all that, we of CIRCA thank YOU, our friends here at this march, for being part of a beautiful and loving nonviolent social movement to achieve true justice . . . for of course it is only through social movements that real social change can happen – not from begging the politicians or corporations to behave more **nicely**.

BUT . . . the CIRCA *Advanced Intelligence Team* has discovered a grave threat to our powerful movement, just as we are all gaining momentum and making our demands for profound change heard. It has come to our attention through deft infiltration, clever clandestinity, and watching the telly that SEVERAL CRIMINAL, ANTISOCIAL ELEMENTS are trying to **HIJACK and CO-OPT** our **MOVEMENT!** They are HERE, AMONGST US, these DANGEROUS EXTREMISTS, and they MAY even be marching in this very demonstration!!! They are **GORDON BROWN (MP)**, the financier of the invasion of Iraq, whose £5 billion War Reserve Fund could, instead, fully immunize EVERY CHILD in the developing world for two years, who has stated explicitly that aid will go down as much as debt is relieved for a net change of ZERO, and that recipient countries will have to restructure their societies to make them even more vulnerable to the market. **HILLARY BENN (MP)** may also be present, wearing his pretty 'development' mask over his true face of privatization and plunder. **BEWARE!** If you see Brown or any of his Errorist henchmen, please report them to the clown patrol nearest to you on the march, and keep your distance.

CIRCA has mobilized to PATROL and PROTECT our movement, so that it is not HIJACKED by these ghoulish buffoons. Our movement's unique accomplishment is the creation of a grassroots politics WITHOUT professional politicians and parliaments (as they have said in Bolivia, Argentine, Ecuador, Chiapas, and beyond, 'Que se vayan todos! (Out with them all!)' We must not let these brown-nosing interlopers mislead us on our own march!

Operation BROWN-NOSE will involve giving hugs to the needy, playing games with all our friends, and other similarly militant activities. We request full cooperation from the public for this operation.

With love, laughter and red-nosed resistance,
Colonel Oftruth
General Confusion
Clandestine Insurgent Rebel Clown Army

The actual press conference involved feeding organic chips to the media (when most of them balked, Colonel Oftruth was sure to hand-feed them until they all relented) and a demonstration of CIRCA's parade skills as an entire battalion marched directly into (and behind, and on top of) a tiny phone booth. CIRCA then demonstrated its *fishing* movement, explained the premise of Operation HAHAHAA (Helping Authorities House Arrest Half Witted Authoritarian Androids), in which CIRCA planned to help the security forces with their ongoing house arrest of the G8's leaders – the ultimate 'dangerous, antisocial, criminal elements' in the Gleneagles resort. CIRCA then read the Operation Brown-Nose communiqué and scattered in all directions. The next day, several papers, including the *Daily Record*, the *Daily Express*, and the *Daily Star*, quoted at varying length the second paragraph of the communiqué, which enunciated a critique of the G8's actual policies and responsibility for the debt of the Global South, along with the disarming CIRCA claim that they would 'amuse, bemuse, but never bruise'.

Press coverage of CIRCA thus served as a partial disruption of the hegemonologue that relentlessly depicted protesters as violent and dangerous. Even in right-wing newspapers that were hostile to the movement's agenda, quotes and photographs of CIRCA members provided a sort of cognitive dissonance to the reactionary storyline. On one such page of the *Daily Mirror*, under the huge headline 'RIOT POLICE... 1 ANARCHISTS...0: G-Hate Aggro Kicks-off as Cops Crush Violent Protesters', and next to a photograph of a beaten protestor, is the image of Trixie kissing the shield of a smiling policeman, with the subtitle of 'KISS AND MAKE UP: Member of the Rebel Clowns plants smacker on cop riot shield'. The *Daily Mail* also showed a picture of Trixie's kiss next to a photograph of a female protestor being arrested by heavily armored police (Moore & McGregor, 2005; Madeley, Macaskill, Tait, & Grant, 2005). These images may at least have disrupted the constant barrage of dehumanizing rhetoric about the protestors, if only from the jarring juxtaposition of the visuals above the denunciatory text. A reader might have wondered, if only for a split second: if these protesters appeared to just be clowning around and kissing shields, why were they being beaten and arrested? In 'The Carnival Turns Into Anarchy' in the *Scotsman*, a reporter noted:

> Earlier in the day, and on a lighter note, the Rebel Clown Army had been detained while putting on their make up and red noses. Police found an artillery of weapons including a feather duster, water pistols and soapy bubbles. About 30 of the clowns – who wore an unusual combination of army combats, neon pink wigs, colanders on their heads and other fluffy accessories – were surrounded by police in Teviot Place. Some clowns ran away giggling but others were forced to wait while they were searched by police. All the clowns insisted on speaking in high-pitched voices, dancing around and making jokes... (Brown, Gray, Howie & McGinty, 2005)

Despite the very real clashes between police and citizens that happened during the Carnival, this sort of press coverage undermined to some degree the rhetoric of the national security state about the threat of violent protestors.

By the end of the campaign, there were some small signs of fraternization between some of the rank-and-file police and some of the clowns, an exchange, a lessening of mutual fear through the visors and the makeup that covered either side's faces. While the clowns were 'directing traffic' at one intersection, a police wagon rolled by the CIRCA 'checkpoint', and some of the police grinningly gave CIRCA the Clown Salute in an appropriately ambivalent gesture of comradely contempt. Elsewhere, one police officer, hearing the clowns do the 'Yes, lets' exercise (in which one clown says 'Let's all do *X*!' and all other clowns in hearing range must say 'YES! Let's all do *X*!', and do whatever *X* is in as enthusiastic and ridiculous a manner as possible), said 'Let's go to the park, because that's where the clowns are going and we have to follow them!' At one point, CIRCA members were once again confronted with a line of stonefaced police, and one of the CIRCA members began ironically thanking the police for supporting global capitalism. The ironic speech went on a bit long, and another clown (Verson) told the police, 'If you smile, I'll tell him to be quiet'. Several of the police immediately put beaming smiles on their faces (Verson). Later, on the highway approaching Gleneagles, a group of rebel clowns came upon four patrolmen guarding a bridge. The clowns asked them if they wanted to play a game, and they agreed. After an explanation of the rules to 'Giants, Wizards and Goblins', the two sides huddled to choose a strategy, then grinningly lined up facing each other. On the count of three, both lines simultaneously aimed their outstretched arms at each other and wiggled their fingers as if casting a spell. The bobbies and the clowns had both chosen the 'Wizard' option. They immediately obeyed the hallowed rule of the game: if both sides choose the same creature, they have to hug. Police-wizards and clown-wizards hugging after all the conflict of the days gone by was a hopeful if bizarre sight (Young, 2005). While this occasional softening of relations may have been denounced amongst some more 'hardcore' and hostile anarchists, others such as founding CIRCA member Jennifer Verson saw it as a historically-proven essential element in destabilizing power and making real social change – establishing a human connection with the rank-and-file police, eroding the mutual and habitual anger and stereotyping, and experimenting with new methods of interaction (Verson, personal communication, 3 July 2005). CIRCA confronts power by playing with it, by refusing the interpolation of conventional power relations as much as possible, at the immediate point of articulation of that power in public space. CIRCA attempts to create a temporary, evanescent, improvised carnival space where the rules are destabilized and new possible relations are suggested and experimented with (Bey, 1985).

These carnival-inspired power-plays can be problematic. While the experience of training and playing with CIRCA, or with carnivalesque protest in general, can be liberating for individual participants, these actions in and of themselves only hint at a better, possible world. Tactical carnival in and of itself does not change the fundamental relations of production or distribution in the greater society. The liberatory spaces it creates are quickly dispersed, either by the force of the state or by the inevitable need of its participants to eventually *get back to work*. Indeed, these spaces, while embracing a carnivalesque egalitarianism, are not equally open to all. The cost of participation is more easily faced by those with the resources to be able to face arrest, and the race or class privilege that lessens the risks and penalties for confrontation with the state (for an account of the community-sustaining clowning amongst working-class African-American youth in Los Angeles, see the movie RIZE, and Shepard, 2005). Different states have different policing policies, and this radically affects the tactics and range of motion of creative protesters. For example, while, in Britain rebel clowns were able, at times, to interact playfully with the police, in New York, where the policing is far rougher, it is necessary to make the fearsome context part of the shtick. CIRCA/NYC members are therefore much more likely to flee helter-skelter at the slightest hint of trouble and hide behind very small objects. While rebel clowning is a joyful way to find courage and to play with power, there is a necessary element of sorrow or even despair mixed into these performances. It is often probably not the best choice for a movement event. However, no single tactic can solve the problems that the global justice movement confronts, and clowning is only one tactic amongst many in the ongoing experiment of resistance. Indeed, the danger of becoming predictable is all the more incentive for tactical carnival's creation of a space for experimentation with new ideas and nonviolent tactics.

In conclusion, scholars of carnival may very well problematize the global justice movement's interpretation of the carnivalesque. They may dispute such an uncritical reading of the works of Bakhtin, and might argue that this uncritical interpretation undermines the movement's actual praxis around the world. However, although some may claim that this reading of carnival is not historicized rigorously enough, or that it is too accepting of Bakhtin's ideas wholesale, it is unquestionable that this idea of carnival is an emergent frame that is inspiring and galvanizing the theory and action of a global, anti-authoritarian, anti-capitalist movement. This global movement is determined to build and sustain its own cultures, in defiance of the homogenizing corporate monoculture that is spreading so rapidly. Part of the development of these oppositional cultures is the development of a new concept of carnival, influenced by older concepts but continually developing its definitions, parameters, and tactics through activist praxis. This movement's project is to reclaim the carnival for its own purposes and agendas, against the society of the spectacle and the hegemonologue, and for sustained, deeply oppositional, creative and egalitarian activism.

References

Ainger, K. (2004). A global carnival of the dispossessed. In E. Yuen, D. Burton-Rose & G. Katsiaficas (Eds.), *Confronting capitalism: Dispatches from a global movement* (pp. 33–35). Brooklyn: Soft Skull Press.

Bakhtin, M.M. (1968). *Rabelais and his world* (H. Iswolsky, Trans.). Cambridge, MA: MIT Press.

Bey, H. (1985). TAZ – temporary autonomous zones. Retrieved September 12, 2005, from http://www.hermetic.com/bey/taz_cont.html

Boal, A. (1985). *Theatre of the oppressed*. New York: TCG.

Bogad, L.M. (2006). A place for protest: The Billionaires for Bush interrupt the hegemonologue. In L. Hill & H. Paris (Eds.), *Place and performance* (pp. 170–178). New York: Palgrave Macmillan.

Brown, A., Gray, L., Howie, M., & McGinty, S. (2005, July 5). The carnival turns into anarchy. *Scotsman*, 4–5.

Carnival for Full Enjoyment. (2005). Flyer/pamphlet. Personal archive of author. Also retrieved November 12, 2005 from www.nodeal.org.uk

CIRCA. (2005a). G8 briefing and operations information. Pamphlet. Personal archive of author.

CIRCA. (2005b). Radical origami audience handout. Heart-shaped flyer. Personal archive of author.

de Certeau, M. (1984). *The practice of everyday life*. Berkeley, CA: University of California Press.

Debord, G. (1995). *The society of the spectacle* (D. Nicholson-Smith, Trans.). New York: Zone Books. (Original work published 1967)

FBI. (2001, May 10). Statement for the record, Louis J. Freeh, Director, Federal Bureau of Investigation, on the threat of terrorism to the United States before the United States Senate Committees on Appropriations, Armed Services, and Select Committee on Intelligence. Retrieved on June 23, 2003, from http://www.fbi.gov/congress/congress01/freeh051001.htm

Fremeaux, I., & Ramsden, H. (2007). We disobey to love: Rebel clowning for social change. In D. Clover & J. Stalker (Eds.) , *The arts and social justice: Re-crafting activist adult education and community leadership* (Section 1). Leicester, UK: Niace Press.

Hill, C. (2002). Levellers and true Levellers. In S. Duncombe (Ed.), *Cultural resistance reader* (pp. 17–34). New York: Verso.

Infernal Noise Brigade. Retrieved on December 3, 2005 from www.infernalnoise.org

Kenney, P. (2002). *A carnival of revolution: Central Europe 1989*. Princeton: Princeton University Press.

Madeley, G., Macaskill, G., Tait, G., & Graham, G. (2005, July 5). The carnival that turned sour. *Daily Mail*, 2.

Moore, R., & McGregor, F. (2005, July 5). Riot police . . . 1 anarchists . . . 0: G-hate aggro kicks-off as cops crush violent protestors. *Daily Mirror*, 4.

Notes From Nowhere. (2003). *We are everywhere: The irresistible rise of global anticapitalism.* New York: Verso.
Rhythms of Resistance. Who we are. Retrieved on November 17, 2005, from http://www.rhythmsofresistance.co.uk/?lid=56
Scott, J.C. (1985). *Weapons of the weak: Everyday forms of peasant resistance.* New Haven: Yale University Press.
Scott, J.C. (1990). *Domination and the arts of resistance: Hidden transcripts.* New Haven: Yale University Press.
Shepard, B. (2005, August 16). The new model army of clowns: From RIZE to the G8 zaps, clowning asserts a radical imagination of resistance. *Monthly Review online.* Retrieved on September 17, 2005, from http://mrzine.monthlyreview.org/shepard160805.html
'We Will...' Handkerchief. From the author's archive.
Young, Z. (aka Private Individual, CIRCA). (2005). Personal video footage of CIRCA actions.

Carnival time in the Kingdom of Coal

Mary T. Hufford

Folklore Graduate Group, University of Pennsylvania, Philadelphia, USA

This essay examines the ideas of carnival and the carnivalesque in relation to a modern social formation, which, cued by the name of its central figure, is referred to in this article as the Kingdom of Coal. Conjured incessantly from disjunct social perspectives on fossil fuel extraction and energy production, the Kingdom of Coal offers a grounded example of what Charles Taylor terms the 'modern social imaginary'. Carnivalesque maskings and unmaskings dramatize a struggle for publicly relevant subjectivity in the United States. Social bodies constructed through such carnivalesque tropes as the grotesque body, the slaying of the king, and (ecologically) gay materiality articulate profoundly different stakes across class lines. Emulating the public space needed for critical reflection on such imaginaries, this essay experimentally brings into dialogue voices kept rigidly separated in channels of environmental decision-making and the national media. Mikhail Bakhtin's concepts of monologic and dialogic modes of communication, and the related concepts of classical and grotesque bodies and ecologies, drawn from his theorizing of carnival, are key.

The Kingdom of Coal as a modern social imaginary

Conjured incessantly from disjunct social perspectives on fossil fuel extraction and energy production, the Kingdom of Coal offers a grounded example of what Charles Taylor terms the 'modern social imaginary' (Taylor, 2004). The metaphor of King Coal puts flesh on the abstract forces shaping society around the production of energy from fossil fuel. I argue that the name, and its aura of uncontested authority and power, persists partly because it points to a paradox, either marvelous or scandalous, depending on one's point of view: a *monarch*, whose ascent to the throne was itself the outcome of a spectacular *coup d'état*, calls the shots in a democracy. The times and spaces of King Coal disrupt modernity as outcroppings of an antiquated social and economic order that is anything but orderly. Mary Harris Jones glimpsed this contradiction in the tent colonies of striking miners nearly a century ago. Years later in her autobiography, she concluded her description of that scene with her famous comment: 'Medieval West Virginia! With its tent colonies on the bleak hills! With its grim men and women! When I get to the other side I shall tell God Almighty about West Virginia!' (Jones, 1969, p. 235).

Mother Jones' assessment has circulated among activists and labor organizers as a means of characterizing the Kingdom. It is the kind of statement that can be misunderstood as a comment on the backwardness of mountaineers, whom the captains of industry have in fact tried to modernize by providing them jobs. This 'politics of time' (Reid, 1972) suffuses a script that produces and reproduces the Kingdom of Coal and its subject positions. The conceit of the Kingdom of Coal illuminates a framework for the script and its performances, making them available for critical reflection in a Bakhtinian light.

Bringing Bakhtin into dialogue with the Kingdom of Coal, I build on foundational discussions underway since the early 1990s among humanists, critical geographers, and literary and social ecologists (Holloway & Kneale, 2000; McDowell, 1996; Murphy, 1995; Bell, 1994; Gardiner, 1993; Folch-Serra, 1990). As Bell observes, though Bakhtin never used the term 'ecology', carnival is clearly a means of recovering alternative, non-official ways of knowing and being-in-relationship that have been subordinated by official discourses with escalating intensity since the time of Rabelais. Though Bakhtin identifies such alternatives as 'folk' or 'popular', he situates them within matrices that are profoundly ecological.

The discourse issuing the Kingdom of Coal bristles with scripts, tropes, and arguments rooted in antiquity, many of them concerned with why and how to structure and maintain (or *not*) the distinction between popular and aristocratic ways of knowing and being. In *Rabelais and His World*, Bakhtin (1984b) traces the history of class distinction as it finds expression along two vectors: classical vs. grotesque representations of the body, and monologic vs. dialogic modes of communicating. The classical body, epitomized by Greek sculpture, privileges symmetry and closure. Orifices in a body that is already finished and perfected are sealed against the world. In contrast, the grotesque body is permeable and open to the world. Productively engaged with other bodies via its orifices, the grotesque body rejects symmetry and hierarchy in favor of growth and reproduction (Stewart, 1984, p. 105). Bakhtin's distinction between monologic and dialogic modes of communicating is analogous. Predicated on an impossibly transcendent, dogmatic, authoritarian subject position, monologism refuses any Other, and assumes the final word. In contrast, dialogism is drenched in 'addressivity', and marked with awareness of its participation within a larger, always unfolding system of meaning-making. Final words are impossible.

The carnivalesque is dialogic. Exuding social space for 'time-out' from the constraints of ordinary living, it exults in the body's openness and membership in an ever-growing, ever-renewing *social* body. Between the historically-contingent social body, and the individual body that is its expression, the world unfolds as the medium and expression of social relations. The logical leap from dialogism to ecology is not a big one. Suggesting that the repudiation of the grotesque itself is 'one of the most powerful cultural forces behind the ecological degradation of the planet', sociologist Michael Bell extrapolates from Bakhtin's theory a taxonomy of interacting ecologies, including a 'grotesque ecology' under assault by a 'bourgeois ecology'. Bell notes that exaggerating the body's separation from nature and even from its own bodily functions marks elites with the distinction needed to maintain class position. And the cost is ecological: 'We cannot see how our efforts at social distinction are devastating other life forms and the resources upon which we all depend, for admitting these connections would undermine the feeling of separation bourgeois life seeks. The problems of social inequality and ecological decline are thus indissolubly linked' (Bell, 1994, p. 82).

If, as Bakhtin wrote, 'To be means to communicate' (Bakhtin, 1984a, p. 287), the carnivalesque replenishes the very foundation of being. Carnivalesque dialogue exudes festive spaces wherein voices can engage in 'free and equal dialogue' (Dorst, 1983, p. 418). However 'free' it may be, carnivalesque dialogue is not formless. It diverts to its own ends the genres of communication, or structures of experience and feeling, that relate us to Others. Our participation in such genres implicates us in a history that transcends what Charles Taylor calls 'secular time'. Communicating, we lend ourselves to the symbolic structures that assemble us in the company of speakers past and future, and catch us up in the sort of 'higher time' that, according to Charles Taylor, has gone missing from modern social imaginaries (Taylor, 2004, p. 186).

Postmodern geographers have found in Bakhtin's dialogics a way to understand how social space and genres of communication are mutually constitutive, and an invitation to explore how Carnival is accessed through everyday speech and practice. 'As the dialogical Other of official culture', write Holloway and Kneale, 'Carnival must always be present; it contaminates the supposedly monological utterances of the powerful. Carnival may be a weakened force, but its currents still run through popular culture. In this sense, we should be looking for elements of everyday life which can become "Carnivalised"... open to the play of dialogue, resisting the "last word"' (Holloway & Kneale, 2000, p. 81).

I argue that what is achieved through carnival is *not* an alternate world, but an alternate, ludic *perspective* that fleetingly exposes and distorts *this* world and the identities of its subjects. This perspective, however fleeting, is gained through practices that carnivalize the Kingdom of Coal. Putting metaphorical flesh on the set of abstractions that hold the Kingdom in place, such practices expose the Kingdom in relation to the subjects it has displaced and denied through monologic techniques. What makes the exposure carnivalesque is a logic that juxtaposes lived reality against its idealized renderings (Bakhtin, 1984a, p. 134), exposing within the pale of a ludic flash the impossible paradoxes and contradictions on which the Kingdom rests, uncrowning and degrading the reigning distinctions between proper moderns and 'others'.

I examine three sites where the carnivalesque is staged. The first is the theater of protest staged before the façades of state and national government. In the genre of the façade ritual, revelers mask themselves in order to unmask the 'true' subjects of the state. A second site is revealed in messages from industry and from the state that respond to the protesters. Issued in the same monologic voice, these messages conjure a perspective for a *corporate* state itself; a schizophrenic perspective that paradoxically both asserts and denies the state and its subjects.

Upstream of these conversations, where headwaters gather at the sites of extraction, a shadow public tracks its standing in a phantom public space. Here the carnivalesque piles up into a viewing position on an encompassing order. This contingent public space of the mountains is hermetically sealed within the absolute private property of absentee owners. This property regime refracts a stigmatized 'hillbilly' image and blocks 'local' access to a national public sphere. Examined together within the context of a distinctly industrial modern social imaginary, the carnivalesque illuminates struggles for subjectivity in different quarters of the Kingdom, while training a droll light on the Kingdom's ideological moorings.

The theater of political protest in a shadow kingdom

Fieldnotes, Charleston, WV April, 1998: I am standing in a crowd on the west side of the West Virginia State Capitol Building in Charleston, at a protest rally against a form of stripmining known as 'mountaintop removal and reclamation'. Activists, mostly from West Virginia, but also from Kentucky, Ohio, Tennessee and Virginia, states most affected by the practice, have staked out a space of protest through an array of visual markers. Tables lining the walkway to seating before the podium feature brochures, bumper stickers, and shocking images of mountaintop removal sites. There is an advertisement for an R-rated movie entitled 'The Rape of the Mountains'. Produced by the St. Louis-based Arch Mineral Corporation, the film stars 'Big John the Dragline', Arch's famed ten-story high earth moving machine with a 103 cubic yard bucket. Slogans proliferate on banners, t-shirts, and posters: 'Land Reform, Not Wrecklamation'; 'Almost Level, West Virginia', 'MTR: A Modest Proposal?' and 'Keep the mountains, Remove Miano' (Michael Miano, a former executive with Pittston Coal, was then head of the West Virginia Division of Environmental Protection). On a lawn to the left of the walkway is a cardboard cemetery, created by West Virginia artist and activist Carol Jackson. The cemetery consists of more than 900 tombstones. On each tombstone is inscribed the name of a tributary now buried under 'valley fill' (industry's term for the 'landform' created by dumping mine spoil into an adjacent stream): 'Drawdy Creek', 'Five Tributaries of Camp Creek', 'Lilly Fork of Elk'. The inscriptions represent more than 1000 miles of headwaters buried under landfills. Yet, in spite of these elegiac displays, the mood is festive and ebullient (Figure 1).

Figure 1. Stream cemetery, by West Virginia artist Carol Jackson, laid out on the west lawn of the West Virginia State Capitol in Charleston. Photo by Mary Hufford.

In performances and speeches at the podium, the event unfolds as carnival, in Bakhtin's terms: 'syncretic pageantry of a ritualistic sort' (Bakhtin, 1984a, p. 122). What makes this pageantry syncretic is the appropriation of official channels of discourse for dual and diametrically opposed messages. West Virginia Secretary of State Ken Hechler, the only state official publicly opposed to mountaintop removal, performs original parodies with his sidekick, an attorney named George Dougherty, whose stage name at these events is 'the Duke of Elkview'. The Duke is dressed in red, white, and blue. Heckler and the Duke a parody of the John Denver song, 'Almost Heaven':

> Almost level, West Virginia,
> Scalped off mountains dumped into our rivers.

Then, exhorting the crowd to 'Picture West Virginia without a King', the leader of West Virginia Citizen Action launches the ritual death of the king and the resurrection of the people. At other demonstrations this ritual has been carried out theatrically, through staged confrontations between Mother Jones and King Coal, but this time it is simply proclaimed. Wrapping up the speeches, Denise Giardina, an award-winning novelist who is running for state governor on the mountain party ticket, declares: 'King Coal is dead! Long live the people of West Virginia!'

Giardina invokes something akin to what Bakhtin terms 'the primary carnivalistic act ... the mock crowning and subsequent decrowning of the carnival king' (Bakhtin, 1984a, p. 124). But here we find the business of carnival reversed. 'King Coal', the pretender to a throne that is not supposed to exist in a democratic society, is *uncrowned*, and the people are crowned in his stead. The body of King Coal is not what the body of the king was in the time of medieval carnival. Anchoring the people as its subjects, the king's body was the public and palpable sign of peoplehood. In the late twentieth century theater of political protest, King Coal is not the people's body but the usurper of territory that grounded 'the people' as subjects of the nation-state. Giardina's proclamation is a pragmatic gesture intended not to renew the reigning order but to *change* the politics. Comaroff and Comaroff define ritual as 'an experimental technology intended to affect the flow of power in the universe' (Comaroff & Comaroff, 1993, p. xxx). A carnivalesque view of the world provides the crucial torque for these experiments (Hufford, 2004).

Since the turn of the nineteenth century, in West Virginia, King Coal's supremacy has been undisputed. Contemporary news headlines animate the monarch and posit his corruption at various levels: 'As Hills Fill Hollows, Some Fight King Coal'; 'King Coal Robbing West Virginia'; 'King Coal: Angel with a Dirty Face'; and even, 'King Coal is Dying'. Reading the papers, we learn that King Coal has a thumb under which he keeps the state of West Virginia. His appetite for coal is boundless and he satisfies his lust in the beds of politicians and the UMW alike. Through strategies of 'double voicing', street theater reproduces and cancels out the channels of power.

Double-voicing and mesalliances

Exercising a kind of sympathetic magic, double-voicing hollows out the genres used for official purposes and fills them with subversive messages. This illicit mixing, which Bakhtin terms 'mesalliance', forces reflection on the means and meanings of cultural production (Bakhtin, 1984a). As we know, 'double-voicing' is what makes parody work. At a demonstration outside of a meeting of A.T. Massey stockholders, Vivian Stockman costumes herself as King Coal, a very obese, suspender snapping, cigar-puffing executive in a top hat. Swaggering around, puffing and snapping, waving money at the crowds, the King inquires loudly: 'Any politicians for sale?' Several years later, when A.T. Massey Coal Company CEO Donald Blankenship pushes the legislature to appoint his nominee for judge, Stockman struts the streets as Blankenship, animating a puppet representing Blankenship's candidate for judge.

At many rallies and demonstrations, musicians and songwriters distribute lyrics to parodies of the state anthem, 'The West Virginia Hills'. The song, written by Ellen King in the 1880s, was set to a rousing tune in 1913 by Henry Engle, and adopted as the official state anthem fifty years later. Every child in West Virginia learns it in school:

> Oh, the West Virginia hills! How majestic and how grand,
> With their summits bathed in glory, Like our Prince Immanuel's Land!
> Is it any wonder then, That my heart with rapture thrills,
> As I stand once more with loved ones On those West Virginia hills?
>
> *Chorus*
> Oh, the hills, beautiful hills, How I love those West Virginia hills!
> If o'er sea o'er land I roam, Still I'll think of happy home,
> And my friends among the West Virginia hills.

Protesters sing a parody with many more verses, composed by Greg Casto in the 1990s, to the same rousing tune. Only the first verse and refrain are included here:

> WEST VIRGINIA HILLS (updated)
> Oh, the West Virginia hills! How much longer will they stand?
> The coal barons want to destroy them for the wealth beneath this land –
> Is it any wonder then, that my heart with sorrow fills,
> As they destroy the place I love – Those West Virginia hills.
>
> *Chorus*
> Oh, the hills, beautiful hills, how I'll miss those West Virginia Hills!
> Is there nothing that can stop them from removing mountaintops,
> Before there are no West Virginia hills?

Through double voicing (the tune, haunted by the original words, recoupled with words that mock them) contradictory messages are brought into dialogic confrontation so that within one utterance, two consciousnesses grapple until one devours the other.

Similarly, activists subvert coal industry messages, packing them with competing messages in an effort to tack onto King Coal's official, monologic discourse a last word that reverses the meaning. Thus, when industry reminds citizens that 'Coal keeps the lights on', protesters create bumper stickers that add: '– in West Virginia funeral homes'. To industry's claim that mountaintop removal will create the flat 'developable land needed to improve West Virginia's economy', Denise Giardina tells the crowd of demonstrators: 'I have changed my mind about Mountaintop Removal Mining. I think I may have misjudged the coal companies. Mountaintop removal does provide some economic opportunities to take advantage of. Why not bottle the water from sludge ponds and sell it as "Arch Mineral Water"?' 'Mineral water' effects a mesalliance of two moments that bourgeois consciousness tries to keep separate: elimination and consumption. Giardina makes of the word 'mineral' an utterance within which two consciousnesses and time-spaces vie for authority. The consciousness that makes West Virginia the site for eliminating coal waste vies with a consciousness that would make West Virginia the site for producing potable water. She also exposes historical memory of a nineteenth century travel capitalism when West Virginia's mineral waters were an economic asset.

Social bodies in conflict

The theater of protest conjures two kinds of social body formed in relation to democratically elected leaders and monarchs, the body of the King, and the body of the democratic state. King Coal symbolically incarnates abstractions of a reigning order that defines and regulates social relations, power relations, cultural institutions, and material practices around the planet. It is he, not the peoples' elected representative, who occupies the monumental architecture that is the emblem of the state. Through the familiar protest tactic of the façade performance, activists force a public dialogue with and scrutiny of the occupants of government buildings, including the West Virginia State Capitol, the US Capitol, the regional headquarters of the US Army Corps of Engineers (which issues the permits for valley fills), and the State Division of Environmental Protection.

Folklorist Dorothy Noyes defines the façade performance as 'an enactment staged by one public, collective face before a façade that symbolizes another public face, in order to make demands of those within, who are in turn expected to respond' (Noyes, 1995, p. 4). Forms of masking and unmasking are used to structure 'a conversation between two public faces; a mask talking to a façade and the façade's response. But behind each face are private agents, each endeavoring to divine the intentions and reactions of the other – to see behind the mask' (Noyes, 1995, p. 7). Protesters focus on a discrepancy between what the state's façade expresses and what it conceals, between an opaque surface and a 'real' interior. Beneath the dome representing the head of state, the façade of the capital building functions as the state's public face. Invoking the *vox populi*, protesters claim to represent the true state locked outside of a capitol which has become, in effect, a sham: a *façade* through which industry speaks; democracy the mask worn by a despot, King Coal. State and industry officials counter by challenging the credentials of activists, who are really 'environmental extremists' posing as citizens.

Playing to the broader community, which, as Noyes observes, 'is often an external audience judging between the two performances', both sides rely on the media, which carries the theater of carnival beyond the site of a particular demonstration. In editorials and press coverage, we find the metaphor of the body extended to the land, which often appears as a body violated, maimed, mutilated, murdered. In editorials of the *Charleston Gazette* mountaintop removal is described as 'scalping the mountains', and 'decapitating the skyline'. The *Washington Spectator* deplores the 'grotesque scars' left by 'arboreal amputation' and the 'massive mauling of the landscape'. Press interviews with environmental advocates amplify the embodiment of landscape. The mining chairwoman for the West Virginia Highlands Conservancy, likens the infilling of feeder streams to 'cutting off the fingertips of the watershed'. The Conservancy's president, compares reclamation to 'a prosthesis worn by an amputee'. And in the reclaimed surfaces of the post-mining landscape – grassy knolls, earthen wedges, industrially-designed drainage systems, impounded slurry, and sediment ponds – secretary of state Ken Hechler discerns yet another mask: 'lipstick on a corpse'.

In the depiction of murdered landscape is an allusion to what Jung calls 'the monologic of violence', which 'ends to foreclose the possibility of any dialogue since it wishes to destroy or kill the other' (Jung, 1990, p. 93). Symptoms of a dysfunctional political system congeal into a diagnosis of the decline of the state itself, and the destruction of its subjects. In 'the mountain state' of West Virginia, where the motto is '*montani semper liberi*', mountaintop removal threatens civic identity. 'What will the mountaineers be when the mountains are gone?' Giardina asks the crowd at the state capitol. 'Flateers?'

What is incarnated in images of the mutilated land? Is it the body politic of the nation state, which defined its subjects in relation to a bounded territory, which included majestic purple mountains? Some theorists argue that organic metaphors for the social body signal the struggle of a new subject to emerge out of the destruction of old forms of domination. Francis Barker writes, 'If the once-full body is so often presented as a shattered wreckage of disarticulated fragments, it is because the disintegration of this world and its signification is already upon it. As modern subjectivity begins to emerge, it turns destructively on that older body from which it struggles to free itself' (Barker, 1996, p. 37). But what new modern subject will emerge from the wreckage of the older subject's body? As protesters struggle to free the state from the body of King Coal, the King struggles equally to shake himself free of the 'subjects' who were once represented in the monarch's body. And what of the mountains? Are they part of the King's new private body or the peoples' new public body? Putting flesh on both subjects, the theater of protest provokes the public conversation it dramatizes.

The corporate state strikes back

Industry and the state issue timely public reminders of West Virginia's indebtedness to coal for its livelihood and identity. The West Virginia Senate passes a resolution in support of the coal industry, recognizing its contribution to the state; West Virginia University institutes a 'Coal Hall of Fame', into which, with much fanfare, industry inducts its first members; the state's Division of Environmental protection dispenses awards to industry for reclamation and 'mountaineer guardianship' (mine safety); the National Park Service launches a Coal Heritage Initiative in the coalfields, where the governor approves $220 million dollars for a new throughway named 'The King Coal Highway and Coalfields Expressway'; the state erects a monument to coal at the Capitol, a bronze figure of a miner mounted on a base with bas relief images of draglines and longwall mining machines – a frank celebration of giant earthmoving technology.

In addition to mounting costly advertising campaigns on television, billboards along interstates, and the Capitol South station of the Washington DC metro, industry stages a counter attack on 'environmental extremists'. In its publication *The West Virginia Coal Bell* the activists are often represented as animals or as women. One *Coal*

Bell cover depicts environmental activists as ostriches with their heads in the ground of a reclaimed mine site, oblivious to flora and fauna flourishing on the surface (i.e. peering behind a mask which is *not* a mask). Another shows a woman from the West Virginia Highlands Conservancy leading inspectors from the federal Office of Surface Mining (shown as hunting dogs) along a cold trail to an abandoned mine to sniff out violations. A hound from the state's Division of Environmental Protection watches in bewilderment. A third cartoon shows the same woman posturing behind the mask of a miner, while a real miner crumples a pink slip in his fist.

The feminizing of those siding with nature, down to the invocation of Mother Jones, and the masculinizing of industry as King Coal summons up the deep history of gender in environmental politics. Allusions to the rape and mutilation of the land recall symbolic operations performed on the early modern social imaginary by Francis Bacon, the 'father of industrial science'. Carolyn Merchant argues that in order to break open a space for an experimental science that could turn nature into a resource for economic production, Bacon had to dismantle the popular ecological principles equating earth with maternity. Conceding that nature was a woman, Bacon averred that she was the *wrong kind* of woman, a 'harlot', in fact, to be 'forced out of her natural state and squeezed and molded by art and the hand of man' (Merchant, 1983, pp. 170–171). A version of Bacon's geomisogyny finds contemporary expression in such comments by industrial spokesmen as: 'Opportunities to reconfigure Mother Earth occur daily across the state'.

The carnivalesque engages, as Bakhtin notes, that which is 'materially huge and cannot be overcome by force'. Activists illuminate coal's ubiquitous social, economic, legal, and political outcroppings: the effects of particulate matter on asthmatic populations, the rising incidence of autism related to mercury emissions, the deterioration of the earth's ozone layer, the destruction of watershed and forest ecosystems, and the impacts on communities in the coalfields from blasting, flooding, and habitat loss. As a spokesman for industry put it, 'Coal's tentacles are everywhere'. With its rigid hierarchies, monological discourse, fixed perspectives and identities, and the cosmic terror evoked by the gigantic and monstrous, the articulation of the Kingdom of Coal, not only by protesters, but by its avowed supporters, forms a fertile breeding ground for the carnivalesque in its industrial guise.

Body, landscape, and industrial gigantism

> You are my creator, but I am your master. (Monster to Victor Frankenstein, Shelley, 2005, p. 87)

Making the body and its parts central to its rituals and imagery, the discourse of protest carnivalizes, puts flesh on, makes visible, the corporate state of West Virginia. Bakhtin pits expressions of the grotesque against the aristocratic inclination toward a classical body that is sealed off and symmetrical, the embodiment of social hierarchy. The image of the body works as the measure of a set of exaggerations on which the Kingdom of Coal rests. The image of this social body is completed in the draglines and heavy equipment, to which devotees and journalists alike refer as 'giants' and 'monsters'.

Into this machinery, the coal industry poured its profits, investing heavily during the 1980s, following the passage of SMCRA (and in anticipation of the 1992 Clean

Air amendments that would increase the value of West Virginia's low sulphur coal). It is this machinery which has enabled industry to surplus its unionized workforce and the expense of its benefits. And it is this machinery that now *necessitates* mountaintop removal and the regulatory changes needed to accommodate the machines. 'The technology has become so vast', said an official of the West Virginia Surface Mining and Reclamation Association, 'The massive nature of this industry has delayed the development [of regulation]... The things that fit the mining practices of 1977 need to be revisited to match the mining practices of today'.

This 'vast' technology now requires industry to bypass the state and federal requirements for community involvement in environmental decision making. Redesigning hundreds of square miles of Central Appalachian terrain became a technical assignment for engineers. The giant technology pressured industry to lobby congress to change the federal definition of 'fill material' in the Clean Water Act so that it can legally dump mine waste into streams.

That the machines are called 'giants' is of more than passing interest here. World mythologies teach us that before there can be a landscape, there have to be giants. Creation myths often detail the titanic struggles that produced the present landscapes, which even in secular terms are littered with the body parts of divinities and giants. In Native American mythologies, land forms cohere as the body parts of divinities, or the trace of their actions, which produced the world and the culture required to inhabit it. A Cherokee creation myth accounts for Appalachian mountains and valleys as the effect of the Great Buzzard, who grew tired while searching for dry land, and where his powerful wings touched the wet earth, they punched it down into valleys, raising the mountains around them (Mooney, 1900) (Figure 2).

However unwittingly, even the landscapes of English speakers are saturated with the traces of giants: the heads and mouths of hollows, spines of ridges, gorges of rivers, feet of mountain, necks of woods, bowels of the earth, and so on. Susan Stewart observes that such references allude to the gigantic that stands at the origin of a public, collective history (Stewart, 1984, p. 71). Landscape features associated with giants become touchstones for narratives of collective origin and unfolding identity in relation to known places, people, and events. Our terms for these features hint at our continuing implication in what Bell calls 'grotesque ecology' (Bell, 1993).

Ecologies are historically contingent and linked to the ideas that relate individual bodies to larger social bodies. Bell's notion of the 'bourgeois ecology' provides a key to understanding the environmental implications of industrial class relations. Even behind landscapes that are the expression of bourgeois ecology there is an etiology of giants. Stewart observes that gigantism in its industrial guise contributes to 'the creation of the new public spaces necessary to class society: the spaces of reproduction and production within which those classes define themselves by means of an exaggeration of boundaries' (Stewart, 1984, p. 83). Not only are the draglines that move mountains referred to by their makers and operators as giants and monsters; they have names, genders, ethnicities, and sexual orientations: Big John, Big Muskie, Sacagawea, and Priscilla (whose 'drag act will be queen of the desert', according to Bucyrus Erie, the manufacturer).

Referred to as kings and queens, the machines are aristocratic, but they are more often referred to as monsters and giants. They walk, they bite, they gobble up dirt, they have 'appetites for huge volumes of earth' (Haddock, 2001, p. 75). Of special interest

Figure 2. The confluence of Drew's Creek and Peachtree Creek, headwaters of the Big Coal River, southern West Virginia. Photo by Lyntha Eiler, American Folklife Center, Library of Congress.

here is the sheer gusto with which the dragline's devotees celebrate its violations of scale. 'Imagine the sight of this monster machine eating up the countryside', writes Keith Haddock, referring to a model released in the 1920s, 'taking eight cubic yards at each bite' (Haddock, 2001, p. 21). (In contrast, contemporary dragline buckets can hold a hundred cubic yards.) And when the machines retire, their devotees raise money to preserve them as industrial heritage sites. Big Muskie's bucket is memorialized as the centerpiece of a campground owned by Appalachian Electric Power in McConnellsville, Ohio. Signs on I-77 advise motorists to get off at exit 25 to pay the bucket a visit (Figure 3).

Lest we read this as the carnivalesque at the service, rather than the critique of those in power, I suspect that Bakhtin would argue that quantitative excess is not enough, especially where there is no hint of ambivalence in the monologic, dogmatic voice of industry, none of the droll perspective of carnival, and no sense of an alternate perspective that is just a ludic glance, word, or gesture away. 'Everything that is truly great', wrote Bakhtin, 'must include an element of laughter' (1986, p. 135). It might be said that the giant machinery destined to destroy the grotesque ecologies of the world shares with *violence* the quality of knowing no laughter (Jung, 1990, p. 93).

Creating the social space of the spectacular technological sublime (Nye, 1994), the giant machinery induces not laughter, but a sense of awe and diminution, as in this description of the GEM ('Giant Earth Mover') of Egypt's open house in 1967: 'The line files past the giant dipper, and there's a stir among the people as they look closely at its hugeness. Families pose in it as they would at Mammoth Cave, and

Figure 3. Big Muskie's Bucket, memorialized at an Appalachian Electric Power campground, McConnellsville, Ohio. Photo by Mary Hufford, Center for Folklore and Ethnography, University of Pennsylvania.

cameras click all around... You can tell people that the house is as big as a three-story, six-family apartment building, but until they roam around it themselves, they cannot feel the sense of magnitude, power, precision, and solidity they experience by seeing and touching... People are swarming all over the machine like Lilliputians' (Haddock, 2001, p. 55).

If the gigantic stands at the origin of a public, collective history, what kind of landscapes may we anticipate from these giants, and what kind of public will take shape around them? Designed from a 'viewshed perspective', the 'landform complexes' left by the giants are monologic and spectacularly empty. But probing more deeply into the language about them we find the bodies. Landform complexes are engineered so that rainwater may accumulate at the 'heads' of valley fills. The overflow is channeled along 'groins' that flank the fill and taper down into ponds at the 'toes'. Decapitated, disemboweled, and sheathed in a 'skin' that exfoliates grasses from the Gobi desert, such a landform annihilates both memory and productivity. What is the public imagined here? 'John Q. Public likes to see grapes', intoned an industry official at a symposium on reclamation.

Gardens and the shadow commonwealth of industrial plantations

Roger Abrahams traces the roots of Caribbean carnival to the slave gardens, which provided a means of access to the marketplace (Abrahams, 1992), a theory that is supported by Eugene Genovese's observation that there was no equivalent in the old

American south to the garden/market relationship, which transformed Caribbean slaves into 'part-time peasants and petty traders' (Genovese, 1974, p. 536). Moreover, the 'yards' over which Jamaican slaves exercised control were social centers that generated supplementary food, and as burial places for family members took on deeper ceremonial meanings. My intention in what follows is not to suggest parallel relationships between gardening and the carnivalesque in the coalfields and gardens and the emergence of carnival in the Caribbean. I want to suggest rather that gardening as it was and is practised in the coalfields expresses a sense of openness to and membership in a constantly becoming world. Such an ecological sensibility, part and parcel of the carnivalesque, is opposed to any ecology that could be termed 'bourgeois'.

In eighteenth-century Virginia, which included all of present-day West Virginia, the slave gardens did afford a measure of autonomy: under the guise of relieving the master's burden of feeding them, slaves won time to provide for themselves, not only by gardening, but hunting, fishing, trapping, foraging, and poaching. Some planters, notes Genovese, recognized the threat to the system posed by such independence, which offered slaves a space for maneuver in which privileges could become rights (Genovese, 1974, p. 540).

The industrial version of the plantation system (engineered in part by descendants of planters who knew how to manage large numbers of workers) inverted this arrangement, wielding ownership of the land rather than ownership of the worker as a means of absolute control. In southern West Virginia, where most of the land had been owned since the late eighteenth century by absentee landlords, settlements of extended families were accustomed to using the mountains surrounding the hollows and valleys as a de facto commons, regardless of ownership (Hufford, 2008). Livestock roamed freely through the woods, and settlers hunted, gathered, and cultivated patches of corn, beans and squash in the mountains. So productive was this corn-woodland-pastureland system that, as David Alan Corbin notes, the early miners tended to work only when they needed cash, and were slow in organizing a union (Corbin, 1981, p. 34). A letter to the *United Mineworkers Journal* in 1896 explains: 'There does not exist the hunger and suffering that is found in... [other coalfields]. Every available spot of ground seems to have received attention from the plow or spade, the houses resemble the home of the market gardener... They raise all the vegetables they require and this assures them that the wolf shall be kept from the door' (as cited in Corbin, 1981, p. 34).

Writing in *The Virginias* in 1885, the English surveyor and speculator David Ansted used the corn-woodland-pastureland system to entice investors to the coalfields, for this system represented a way in which workers might feed themselves, saving operators the trouble and expense. 'By offering allotments of moderate extent on easy terms, reserving the minerals', wrote Ansted, 'population could be secured and the whole district improved. The reserved allotments would soon increase in value. The cost of placing emigrants on these lands, from Liverpool, would not exceed $45 per head' (Ansted, 1885, p. 129).

Clearcutting the virgin forests and quadrupling the population with immigrant workers, industrialists rapidly depleted the surplus value that subsistence farmers had banked on. This extractive economy begins with the impoundment of collective becoming. While everything else in the local ecology could continue to circulate, the land companies froze the timber and minerals, fixing their identities as 'privately owned resources'. The crippling effects would not be realized locally for decades, partly because the coal companies encouraged gardening in the company towns, even inaugurating garden competitions in an effort to show the world that they were good stewards of the labor force.

The relationship between gardens and the marketplace was radically transformed under the property regime of coal. Laborers were free by definition, but the marketplace of the company store was not. (The company store in fact served as an economic buffer for coal companies during times of bust.) Surplus value for the miners entered in the form of the gift resources 'imported' from mountain streams, forests, and soils. The right to hunt, garden, fish, and forage on the company's land allowed for the continuation of a barter and borrow economy among workers, whose families traded produce from woods and garden (including moonshine). But while the marketplace offered the prospect of a different identity and possible freedom for slaves in the Caribbean, in the coalfields, the company store was a means of trapping labor, keeping it available when work was not. Corbin argues that for miners with a plot of ground to work, gardening offered a way to stay out of debt, and to survive without pay during strikes over contracts with industry (Corbin, 1981, p. 35). Like the slaves who tilled their garden patches by moonlight, miners were known to plow newground under the stars in hopes of breaking even.

The social body that formed around coal production in this era was a Fordist one, an industrial mutation of the king's body. 'Henry Ford said that a few men work from the neck up, but most work from the neck down', a miner remarked to me one night in the Sundial Tavern. 'What do you think of that?' After workers won the right to bargain collectively in 1933, industry negotiated with the UMW, promising to set up a pension fund, so long as miners would not protest the loss of jobs due to technological advancement (Couto, 1993). When the mines mechanized in the mid-twentieth century, 250,000 miners left the Appalachian coalfields for jobs in midwestern factory towns. With the development of giant earthmoving technology in the late twentieth century, more coal is produced than ever before, and with fewer workers. Couto observes that the acronym for the governing principle, 'return on investment', is the French word ROI, reminding us that Coal is still King (Couto, 1993).

Land companies continue to occupy a unique and surprisingly invisible position in this economy. In southern West Virginia where 90 per cent of the land is owned by absentee companies, the parcels are unusually large – tens of thousands of acres. In this system, a land company's neighbors are not the neighbors of the tenants who occupy land company property, but the adjacent land companies, those who share in ROI. For example, the major landholders on the Marsh and Clear Forks of Coal River, the Rowland, Berwind and Dickinson land companies, possess adjoining properties and maintain offices near to each other in Charleston. This makes them neighbors. All three are listed among the top fifty land companies (Appalachian Land Ownership Task Force, 1983) and they don't occupy what they own. They lease most of what they own to coal and gas companies. 'That's the reason this place could never grow up', one man on Coal River told me. 'The land companies will lease to you, but they won't sell' (personal communication, 21 May 1994).

The encompassing of settlements in the mountains was accomplished through a maneuver to which state and industry documents refer as the 'severance of ownership from occupancy'. The resulting arrangement, which Kathleen Stewart

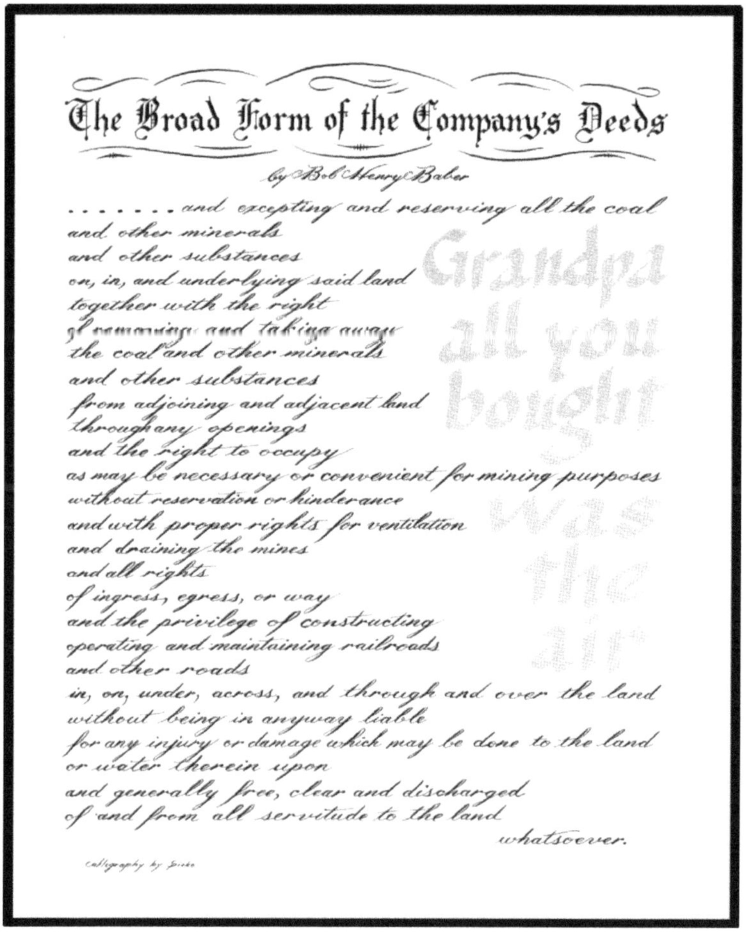

Figure 4. 'The Broad Form of the Company's Deeds', a parody by West Virginia poet Bob Henry Babar, inspired by the deed to his grandfather's farm. Courtesy of Bob Henry Babar.

terms 'double occupancy' (Stewart, 1996), continues to generate the wealth that sustains the political power and technological control of coal. Double occupancy and its contradictions give rise to forms of carnivalesque expression that are markedly different from the carnivalesque expressions of political activism. Both sets of expression are directed at the same formation of power, and what is at stake might be thought of as subject position, but in the coalfields, where the domination of industry has been felt as a formidable force for more than a century, the struggle is not so much to occupy civic space as to occupy any space at all (see Reid & Taylor, 2010) (Figure 4).

For people living near the mines, the condition of double occupancy gathers into a high pressure surround that can be fragmented and momentarily lifted through carnivalesque practices, images, perspectives, and laughter. It is not the words of industry that are subverted, but the landscapes themselves that are doubled into simultaneous media for 'official' and carnivalesque messages.

Laughter at the foundations of becoming

> Prepare for mirth, for mirth becomes a feast. (Simonides, in Shakespeare's *Pericles, Prince of Tyre*, Act II, Scene III)

Bakhtin argues that carnivalesque laughter is a weapon against cosmic terror, which he defines as 'the fear of that which is materially huge and cannot be overcome by force'. The condition of double occupancy makes the coalfields paradoxically both commons and commodity. Carnivalesque images and practices become strategies for the fleeting recuperation of place by degrading and uncrowning the ever-present cosmic terror of annihilation.

Imagine knowing that water is building up in sediment ponds and subterranean passages above your home and not knowing when or where the next 'blowout' or flood may occur, not knowing what will become of that water when the mountains are blasted apart for coal, knowing that if you complain you'll lose your job or your lease on the company's land. Imagine that your house, two rooms wide, two rooms deep, flush with the side of the road, is where the Peabody coal company moved you when it needed the hollow that your family farmed for five generations for another purpose, namely a place to put the sludge from the coal cleaning process, in compliance with federal clean air regulations. Now your children, who attend the elementary school just beneath the 800 foot high impoundment, exhibit chronic respiratory problems and other ailments. Your house is coated with black dust billowing from coal trucks barreling down the two-lane 24-7, drowning out the television in your living room and the conversations on your porch. Your house is downstream from the sludge dam, so whenever it rains, you are checking the level and color of the water in the creek running behind your house, because clear-cutting has exacerbated flooding, and sludge dams have been known to leak and to burst, carrying vehicles far downstream and drowning neighbors.

In this case the giant that terrifies is a corporate one, impervious to the effects of its actions on the unincorporated inhabitants of its property, effects that include dust from coal trucks, foundations cracked and wells dried up from blasting, flooding and drownings linked with clear cutting and impoundments. As a merchant in Stickney put it, speaking of the sludge ponds where billions of gallons of waste from the cleaning of coal is stored upstream of thousands of people: 'We fear the river above more than the river below' (personal communication, 27 September 1994).

Mountaintop removal mining is rationalized through operations that belittle the culture of the mountains, stigmatizing 'mountain ways' of speaking and relating to the land as signs of economic backwardness and cultural deprivation. A culture that industry reduces partly by making itself materially huge is officially contained by the state, often romantically, in themed worlds of coal heritage and pre-industrial domesticity: marketing craft, music, and storytelling as 'tradition', and producing new landscapes as 'viewsheds'. Fixing Appalachian culture in the past, these sites contribute to perceptions of Appalachia as a site of 'permanent outdatedness' (Taussig, 1993). As Kathleen Stewart describes it, an Appalachia outside of the enclosures anchors a bourgeois imaginary of sagging shacks, hand-me downs, trash, and an excess of stray dogs. To occupy this landscape from within is thus to experience the tight squeeze of an encompassing order (Stewart, 1996).

Carnivalesque laughter emerges as a mode of exceeding this tight cultural, economic, and geographic allocation. The limitations are those of fixed identities imposed on people, resources, and commodities. Outgrowing or eluding the felt boundaries of an alien intent to become something else requires excess that is qualitative, not merely quantitative (Bakhtin, 1984b, p. 308).

On the surface, strategies of qualitative excess may simply look like recycling, but in the aggregate they form thresholds to a prized vantage point on the world in its droll aspect. Augur bits from the mines become fence posts. Roof bolts from the mines become tomato stakes. Engine blocks are redeployed as supports for a new log cabin. A bus is refashioned into a pick-up truck. A stop sign patches the bottom of a wheelbarrow. The oak timbers that bridged the Coal River at Edwight are reconstituted as a log cabin. 'You can live in it or cross the river in it', quipped a neighbor who lived nearby. The man who built the cabin admitted, 'A lot of trucks have driven over this house' (personal communications, 1 February 1996).

One of the forms taken by qualitative excess is telling 'lies', a widespread genre of tale-telling associated with male workers (Stewart, 1984, p. 98). On Coal River, exaggeration and hyperbole centered on known people and places proliferate along multiple channels. In the Sundial Tavern, Kenny Pettry, the 'mayor' of Sundial, displays the gifts sent to him by Charles Cantley, the 'mayor' of Dry Creek: a 'Sundial Tongue Depressor' made from a hide stretcher is for 'the big mouths of Sundial'; a Liar's license on the wall certifies the Pettry's capacity for truth-stretching; a 'double-barrelled slingshot' comprises a very large brassiere affixed to a pronged stick; a fly swatter the size of a window screen stands ready to do battle with the huge flies of Sundial; an empty key ring hanging on a hook bears a tag that says, 'Keys to the city of Dry Creek'. Such artifacts display the interest of the grotesque in the body's productive parts and products – the mouth of the liar, the volume of waste that must attract gargantuan flies, the female undergarment that morphs into a weapon – all of these outgrow themselves, transgress their boundaries, and become something else.

The isolation, exaggeration, and display of reproductive organs and orifices shifts the focus, as Susan Stewart observes, from verticality, symmetry, and hierarchy to production (Stewart, 1984, p. 107). Similarly, the landscape on Coal River grows boisterously productive in stories that amplify its reproductive parts. Bud Boggess tells of how he tried to cut down a sapling up on Hazy Creek and when berries the size of tomatoes rain down on him he realizes that what he mistook for a buckeye is actually a giant ginseng plant (berries normally the size of peas). Another man swears he lived for an entire winter on a potato the size of a small truck. 'That's a fellow we worked with', said Elbert Webb, 'He used to get in with a big crowd down there and he'd tell that. He said that potato was so big that he just left a pole axe laying beside of it, he couldn't get it out of the ground' (tape-recorded conversation, 22 November 1997). As with the Rabelaisian grotesque, the exaggerated reproductive organs display parts of a landscape of becoming, known as a body from within – collectively navigated through recourse to what is the equivalent of a collective, or social, body image: a shared imaginary conjured through narrative (Hufford, 2003).

Laughter as spatial practice

> We'd set around on that old slate dump, build a fire, roast hot dogs, talk trash, tell lies and stuff, just to get the laughs. (Ben Clay, Stickney, West Virginia)

Such forms of exaggeration erect and sustain an image of what Bakhtin calls 'the gay material bodily cosmos, ever-growing and self-renewing' (Bakhtin, 1984b, p. 340). Narrating it from within, people rebuild the world in its droll aspect, embodying, degrading, and uncrowning the catastrophic assault of industrial and city ways on the ways of the country, of which mountaintop removal is only one manifestation. Surprised by the world in this aspect, we cannot help but laugh. And here I want consider how laughter functions as a spatial practice, a means of producing and occupying social space.

Psychoanalyst Paul Schilder assigns to laughter the function of breaking down and rebuilding the individual body image (Schilder, 1950). On Coal River and elsewhere, language about laughter registers the impact of laughter on bodies. Stories themselves are materialized as entities that can be spun, laid down, traded, swapped, handled – 'I can't *touch* nary a one you tell', Ben Clay tells Elbert Webb on being told it's his turn to tell a tale. Think of how stories can 'tickle' or 'slay' us. Laughter has a destructive force. We 'shake', 'bust a gut', 'split our sides', 'roll on the ground', 'convulse' and 'die' with laughter. Unlike what Bakhtin calls 'reduced laughter' (the sardonic snicker or polite chuckle that does not get below the neck), carnivalesque laughter (the guffaw, the roar, the belly thumper, the thigh slapper) impacts the whole body. Clearly, if laughter is a weapon, the body is its target. Fragmenting the body image, laughter transiently destroys the constraint we effect for our bodies, a constraint that is historically and culturally contingent, articulated in the grotesque image of the Fordist body: a corporate head, working from the neck up, in charge of bodies working from the neck down.

Ben Clay and Elbert Webb pondered the necessity of laughter for survival in the industrial workplace.

> 'Buddy, I'll tell you', said Elbert, recalling the years he spent working in a sawmill, 'Me and Ben's had some times, when we used to work, laughed all day long, working ten, twelve hours a day... It went all day long, didn't it, Ben?'
>
> 'That's what made your day', said Ben.
>
> 'Yeah, you worked so hard – '
>
> 'You'd stand there –'
>
> 'It kept us a-going'.
>
> 'You'd just reach one board after another, just all day long'.
>
> 'And Hazy's just like a chimney a-drawin', said Elbert. 'Summertime, it was so hot you just couldn't hardly stand it, wintertime you just about froze to death, didn't you?'
>
> 'Oh gosh', said Ben. 'Seemed like that cold air was blowing out of there of a wintertime, and seemed like the hot air was blowin out in the summer'.
>
> 'It was set up on this old red dog –'
>
> 'Old burnt slate – '
>
> 'And you could just feel the heat of that. My feet would get so hot I couldn't stand it, through my boots, where they had that lumber stacked'. (tape-recorded conversation, 22 November 1997)

Staging the stories told to 'get the laughs', Ben and Elbert invoke the monotonous routine and physical discomfort of a Fordist workspace, which they humanized through laughter. They likened the stories they told that evening to the tales they told years earlier while working together in the sawmill. To situate this moment in the history of laughter (the history in which Bakhtin situates Rabelais), I want to look further at a set of images that we might consider to be industrial capitalist grotesques. As Carnivalesque images, they 'turn their subject into flesh, dragging high culture down to the sphere of earth and body in their indissoluble unity' (Holloway & Kneale, 2000, p. 80).

Images of excessive lack in a series of 'hard times' stories graphically explore the absurd logistics of consuming nothing. In contrast to the images of the peoples' ever-growing, feasting body that Bakhtin associates with Rabelaisian laughter, these images amplify deprivation: the couple so hard up they had to share the same false teeth, the family so hard up they allowed each child to swallow a single piece of meat on a string long enough to digest a few molecules before withdrawing the meat for the next child, the man so hard up that he fed his horse dried apples for breakfast, water for lunch, and 'for supper he'd just let it swell'. Set in the particular spaces of a place that, as one man put it, 'could never grow up', these images inflate the condition of lack to preposterous dimensions. In this imaginary positioned at the bottom of the chain of extraction driven by Capitalism's unlimited appetite for growth we find bodies engaged in the paradoxical task of replenishing themselves on nothing, of growing a society where it is not possible even to eat together. In a place that can't grow up, only nothing can exceed itself. The parodoxical task of making *nothing* exceed itself is accomplished through the lie itself, which, once digested, produces culturally regenerative laughter: the feast that is missing in the stories.

And feasting is the point. Not only do stories induce laughter that assails the Fordist body image, but our language about stories suggests they nurture the alternative. Stories and their images can be ruminated, chewed on, and swallowed, 'hook, line, and sinker'. Eating, notes Bakhtin, is a 'significant manifestation of the grotesque body', which is distinguished by 'its open unfinished nature, its interaction with the world'. Coal River is a place doubly caught simultaneously at the beginning of the chain of production, and the end of the chain of consumption. What happens in between? The answer to this depends on whether you are caught up in a 'bourgeois' ecology or a 'grotesque' one (Bell, 1994).

In a series of stories about the impossibility of eliminating things that are used up, Ben and Elbert ponder what it means to be consumers and producers on Coal River. The theme of eking more utility out of what's already been used up is central to a host of images that bring the ideology of the human as a closed off, completed, atomized individual into dialogic confrontation with the ideal of the human as part of a dynamic, continuously dying, boisterously regenerating, productive landscape.

A hillbilly masque and the negative dialectic

When such carnivalesque humor makes it into the public eye, it often takes the reduced and caricatured form of the hillbilly stereotype: Snuffy Smith, the Dukes of Hazard, or the Beverly Hillbillies (Just imagine what it would be like if mountain people actually owned their minerals!). The stereotypes laminate into a thick mask conferred by an irreversible gaze (Aching, this volume).

Inhabiting the hillbilly stereotype through a series of videotaped vignettes on Coal River becomes a way of building a collectively shared body image. Created by Arnold Honaker, of Pettry Bottom, in the genre of the documentary interview, the videos feature friends and neighbors wearing hillbilly hats out in the woods hunting, gathering, making moonshine, and doing battle with a creature called 'the Bradley Mountain Monster'. The videos circulate among members of the community, eliciting robust laughter among viewers with whom I watched the productions. In one interview entitled 'The New Edition', Ben Clay dons a garish hillbilly mask and plays the part of a locally known man whose house collapses from neglect. Arnold Honaker asks Ben how his garden is doing.

> 'I got lettuce and onions coming up', says Ben, his voice muffled behind the mask.
> 'You got plenty of sugar watermelons?' inquires Arnold.
> 'I believe I will', Ben predicts. 'You know that dirt right there where my house fell?'
> 'Yeah', says Arnold. 'I remember that'.
> 'Well that corner post there leaned over, and it pushed that dirt over, and there's nothing but black dirt come out from under that house when it fell. All you've got to do is just throw the seeds out there, they just come up everywhere'.
> 'That would be good dirt, wouldn't it', marvels Arnold.
> 'That's good dirt', affirms Ben. 'Just take a finger and stick that seed in there and it'll come up. That dirt's rich!'
> 'Probably all them boards rotting, wasn't it?' speculates Arnold.
> 'Old shoes and stuff'.
> 'Inner tubes'.
> 'Inner tubes, old pants and stuff, laying around. That's what makes your ground rich'. (Tape-recorded screening, 25 April 1999)

Bakhtin theorizes that chronotopes, which define genres, display concepts of the relation between space and time (1981, p. 85). The genre that Ben calls 'Appalachian stories' is set in the time and space of the nation's poverty region. The interviewer is eliciting details of the life circumstances of the hillbilly. Then, with little prompting from the interviewer, the conversation spirals off into a full-blown lie. Suddenly we are in the space, not of lack, but of excess, wherein the collapse of a house unleashes dirt fertile enough to produce watermelons (also, we might note, with little prompting.) How did this dirt get so rich? By dint of the waste products of civilization, inert, non-biodegradable materials combine with Appalachian dirt to generate rich ground, or what Bakhtin would call 'gay matter'.

In the abrupt shift from the genre of the Appalachian story to the lie we become aware of a second consciousness vying for authority within each utterance. Behind the mask of a hillbilly convinced that the waste products of civilization will recycle into prosperity, there is a consciousness that polyester pants, plastic shoe soles, and rubber tires do not enrich the ground. The mask materializes the mask of the stereotype, which Aching (this volume) notes is conferred on others by a reigning order. The actor behind the mask takes aim at the surface meaning of both the mask and its utterances. In the carnivalesque time of the dialogue, the elements of shoes, inner tubes, and pants transgress their limits, growing qualitatively large as fertilizer. In the time of the production, a new and different consciousness exceeds the hillbilly stereotype marked by the mask. The relationship between trashy yards and trashy people is mocked, while the elevating, romantic notion of the garden as a route to prosperity is uncrowned and degraded. The lived reality of a neighbor's trash-strewn lot, juxtaposed against an ideal of prosperity as the accumulation of material goods, unleashes big laughter, trampling into the dirt the ideology of progress and the social distinctions in which that ideology is grounded (Hufford, 2002).

Laughter literally creates social space. Inflated to bursting point with the 'leaven of laughter', bodies give way under the impact of images and ideas. Laughter shatters the composure of a body image produced and reproduced under conditions of double occupancy, conditions that have for a century inhibited growth and prosperity on Coal River and cultural visibility and participation in a broad public sphere. Fragmenting the body and its image, this laughter registers what Kathleen Stewart describes as a 'satisfaction that comes with fragmenting a felt enclosure which excludes a local real' (Stewart, 1996, p. 128). Separating the self from its stereotypical depiction, the hillbilly mask fragments a felt enclosure, thereby generating new social space. Moreover, if spatial practice is 'to be other and to move toward the other' (de Certeau, 1984, p. 110), laughter – contagious laughter of the sort that ripples involuntarily through a social assemblage – is a hierarchy-levelling spatial practice that dissolves the boundary between self and environment, and self and other, (including, I might add, *animal* others, who, like practitioners of Big Laughter, may 'hoot', 'howl', 'cackle', and 'roar'). Registering, in Taussig's words, 'a tremor in cultural identity and . . . in the security of Being itself', laughter not only assails, it destroys and rebuilds an image of humanity at home in the landscapes of Coal River.

The mountain state in the history of carnival

When John Denver's song, 'Almost Heaven, West Virginia', hit the airwaves in the 1970s, people of West Virginia were delighted enough to adopt the song as an unofficial state anthem. In the early 1990s a *Washington Post* columnist suggested a parodic alternate: 'Almost Haiti, West Virginia', which drew angry responses from West Virginians. The parody, however, suggests a more than trivial connection between the economic histories of the Caribbean island and the mountain state, a connection that Appalachian scholars explored in the 1970s and 80s as the 'internal colony' explanation for poverty levels in the mountain state. However, if the coalfields are a colonized region, the colony is a successor to, not a contemporary of, the colonial spaces of cotton, tobacco, and sugar plantations. *Mutatis mutandi*, the carnivalesque in this kingdom appears in a historically contingent, industrial capitalist guise.

This essay has examined carnivalesque practices staged from different and interlocking positions along Coal's chain of extraction which is class (B. Taylor, 1992). From different quarters of the Kingdom, carnivalesque practices are engaged in the production of political, social, and ontological space. In its varied forms, carnivalesque expression dramatizes an unfolding struggle for publicly relevant subjectivity in the United States. At the top of the chain of extraction, forms of masking and unmasking are used to reinforce the identities and non-identities foundational to the kingdom. The theater of protest negates King Coal's dialectic of identity to create an alternative political space, using forms of masking and unmasking are used to dismantle identities that threaten civic being. In the spaces of unincorporated communities at the headwaters, where coal is extracted, carnivalesque forms of expression uncrown and degrade the inert materiality of bourgeois ideology, prying open the ontological space needed for becoming and its humanizing ecologies. In the aggregate these struggles raise profound and disturbing questions about where and how we locate the agency to shape human and political being – and becoming – in the twenty-first century. If King Coal names the successor to the throne vacated by King Cotton, we would do well to ask what rough monarch slouches toward a kingdom of renewable energy (of corn, of geysers, of wind, of sun) to take over the reins of ROI?

Acknowledgements

This essay has evolved over years of conversation and presentation in a variety of settings. It has benefited from readings and comments by Katharine Young, Dorothy Noyes, Robert Cantwell, Roger Abrahams, Jack Santino, Ana Cara, Betsy Taylor and Herbert Reid, and came to the present form as a result of intensive colloquy with the authors of other essays in this volume, to whom I am especially indebted. I also wish to acknowledge the generous support of the Obermann Institute for Advanced Studies, and the encouragement of Jay Semel and Micheline Crichlow, the John Simon Guggenheim foundation for a fellowship that supported ethnographic research in the Coal River Valley from 1998–1999, and the Ford Foundation for a grant, administered through the Robert Wagner School of Public Service at New York University, that supported a collaborative ethnography of the Ohio Valley Environmental Coalition's practice of leadership for a changing world, and the many insights shared by its leaders, Dianne Bady, Laura Forman (who died at a protest in front of the Army Corps of Engineers regional offices in Huntington, in 2001), Janet Fout, and Vivian Stockman, and by Corinna McMacken a graduate student and co-researcher on that project. Most of all I am grateful to people living in the Coal River Valley whose voices are heard in this article, most of them pseudonymously.

References

Abrahams, R. (1992). *Singing the master: The emergence of African American culture in the plantation south.* New York: Penguin Books.
Ansted, D.T. (1885). Coal and coke: The Cabin Creek Coal Company. *The Virginias: A Mining, Industrial, and Scientific Journal*, *6*, 129.
Appalachian Land Ownership Task Force. (1983). *Who owns Appalachia? Land ownership and its impact.* Lexington: University Press of Kentucky.
Bakhtin, M.M. (1981). *The dialogic imagination.* Austin: University of Texas.
Bakhtin, M.M. (1984a). *Problems of Dostoevsky's poetics.* Minneapolis: University of Minnesota Press.
Bakhtin, M.M. (1984b). *Rabelais and his world.* Bloomington: Indiana University Press.
Bakhtin, M.M. (1986). *Speech genres and other late essays.* Austin: University of Texas.
Barker, F. (1996). *The tremulous private body: Essays on subjection.* Ann Arbor: University of Michigan Press.
Bell, M.M. (1994). Deep fecology: Mikhail Bakhtin and the call of nature. *Capitalism, Nature, Socialism*, *5*, 65–84.
Comaroff, J., & Comaroff, J. (1993). *Modernity and its malcontents: Ritual and power in postcolonial Africa.* Chicago: University of Chicago Press.
Corbin, D.A. (1981). *Life, work, and rebellion in the coal fields: The Southern West Virginia miners 1880–1922.* Urbana: University of Illinois Press.
Couto, R. (1993). The memory of miners and the conscience of capital: Coal miners' strikes as free spaces. In S. Fisher (Ed.), *Fighting back in Appalachia: Traditions of resistance and change* (pp. 165–194). Philadelphia: Temple University Press.
de Certeau, M. (1984). *The practice of everyday life.* Berkeley: University of California Press.
Dorst, J. (1983). Neck-riddle as a dialogue of genres: Applying Bakhtin's genre theory. *Journal of American Folklore*, *96*, 413.
Folch-Serra, M. (1990). Place, voice, space: Mikhail Bakhtin's dialogical landscape. *Planning and Environment D: Society and Space*, *8*, 255–274.
Gardiner, M. (1993). Ecology and carnival: Traces of a 'green' social theory in the writings of M.M. Bakhtin. *Theory and Society*, *22*, 765–812.
Genovese, E. (1974). *Roll, Jordan, roll: The world the slaves made.* New York: Pantheon Books.
Haddock, K. (2001). *Extreme mining machines: Stripping shovels and walking draglines.* Osceola, WI: Motor Books International Publishing Company.

Holloway, J., & Kneale, J. (2000). Mikhail Bakhtin: The dialogics of space. In M. Crang & N. Thrift (Eds.), *Thinking space* (pp. 71–88). New York: Routledge.

Hufford, M. (2002). Interrupting the monologue: Folklore, ethnography, and critical regionalism. *Journal of Appalachian Studies, 8, 62–78.*

Hufford, M. (2003). Knowing ginseng: The social life of an Appalachian root. *Cahiers de Litterature Orale, 53–54*, 265–291.

Hufford, M. (2004). *Waging democracy in the Kingdom of Coal: OVEC and the struggle for environmental justice in Central Appalachia.* Philadelphia and New York: Center for Folklore and Ethnography, University of Pennsylvania and Robert F. Wagner School of Public Service, New York University. Available at http://www.sas.upenn.edu/folklore/center/waging_democracy2.pdf

Hufford, M. (2008). *Ethnographic overview and assessment of the New River Gorge National River and the Gauley River National Recreation Area.* Boston and Philadelphia: National Park Service, Mid-Atlantic Regional Office and Center for Folklore and Ethnography, University of Pennsylvania. Available at http://www.sas.upenn.edu/folklore/center/NGR_report_final.pdf

Jones, M.H. (1969). *The autobiography of Mother Jones.* New York: Arno.

Jung, H.Y. (1990). Mikhail Bakhtin's body politic: A phenomenological dialogics. *Man and World, 23*, 85–99.

McDowell, M.J. (1996). The Bakhtinian road to ecological insight. In C. Glottfelty & H. Fromm (Eds.), *The ecocriticism reader: Landmarks in literary ecology* (pp. 371–391). Athens, GA: University of Georgia Press.

Merchant, C. (1983). *The death of nature: Women, ecology, and the scientific revolution.* New York: Harper.

Mooney, J. (1900). The Cherokee river cult. *Journal of American Folklore, 13*, 1–10.

Murphy, P.D. (1995). *Literature, nature, and other: Ecofeminist critiques.* Albany: State University of New York Press.

Noyes, D. (1995). Façade performance in Catalonia: Display, respect, reclamation, refusal. *Southern Folklore, 52*(2), 97–110.

Nye, D. (1994). *The American Technological Sublime.* Cambridge, MA: MIT Press.

Reid, H.G. (1972). The politics of time: Conflicting philosophical perspectives and trends. *The Human Context, 4*, 456–483.

Reid, H.G., & Taylor, B. (2010). *Recovering the commons: Democracy, place, and global justice.* Champaign-Urbana: University of Illinois Press.

Schilder, P. (1950). *The image and appearance of the human body.* New York: John Wiley & Sons.

Shelley, M.W. (2005). *Frankenstein.* Stilwell, Kansas.

Stewart, K. (1996). *A space on the side of the road: Cultural poetics in an 'other' America.* Princeton: Princeton University Press.

Stewart, S. (1984). *On longing: Narratives of the miniature, the gigantic, the souvenir, and the collection.* Baltimore: Johns Hopkins University.

Taussig, M. (1993). *Mimesis and alterity: A particular history of the senses.* New York: Routledge.

Taylor, B. (1992). *Drinking from the rock: Self and story in Appalachian coal mining communities.* Ph.D. Dissertation, University of Michigan.

Taylor, C. (2004). *Modern social imaginaries.* Durham, NC: Duke University Press.

Index

Page numbers in **Bold** represent figures.